Computational Quranic Studies:

Structural, Mathematical, and Cognitive Modeling

Ali Naqvi
American Institute of Artificial Intelligence, USA

Nurullah Naqvi
American Institute of Artificial Intelligence, USA

IGI Global
Scientific Publishing
Publishing Tomorrow's Research Today

Vice President of Editorial	Melissa Wagner
Director of Acquisitions	Mikaela Felty
Director of Book Development	Jocelynn Hessler
Production Manager	Mike Brehm
Cover Design	Jose Rosado

Published in the United States of America by
IGI Global Scientific Publishing
701 East Chocolate Avenue
Hershey, PA, 17033, USA
Tel: 717-533-8845 | Fax: 717-533-7115
Website: https://www.igi-global.com E-mail: cust@igi-global.com

Library of Congress Cataloging-in-Publication Data

Names: Naqvi, Ali S., 1968- author | Naqvi, Nurullah, 1998- author
Title: Computational Quranic studies : structural, mathematical, and cognitive modeling / Ali Naqvi, Nurullah Naqvi.
Description: Hershey, PA : IGI Global Scientific Publishing, [2027] | Includes bibliographical references and index. | Summary: "This book aims to explore the systematically modeled representation of the Quran using advanced computational techniques"-- Provided by publisher.
Identifiers: LCCN 2026003246 (print) | LCCN 2026003247 (ebook) | ISBN 9798337369402 hardcover | ISBN 9798337369419 paperback | ISBN 9798337369426 ebook
Subjects: LCSH: Qur'ān.--Criticism, interpretation, etc.--Data processing | Computational religious studies | Artificial intelligence--Religious aspects--Islam
Classification: LCC BP130.38 .N37 2026 (print) | LCC BP130.38 (ebook)
LC record available at https://lccn.loc.gov/2026003246
LC ebook record available at https://lccn.loc.gov/2026003247

British Cataloguing in Publication Data
A Cataloguing in Publication record for this book is available from the British Library.

To Imam ʿAlī ibn Abī ālib (a),

Gate of the City of Knowledge, whose words gave reason its dignity and justice its measure.

May this humble effort to read revelation in the language of our age remain faithful to the light he carried.

Table of Contents

Preface

Before Sir Isaac Newton became the immortal mind of the *Principia*, before gravity made him the emblem of modern science, he spent long, silent years pursuing another problem he believed was just as profound: the hidden order of scripture. To Newton, nature and revelation were not separate kingdoms. They were two books written by the same Author, and both demanded disciplined reason. So, the man who measured the motion of planets also turned his mathematical habits toward the Bible: chronology, proportion, pattern, sequence, symbol, and rule. He did not model scripture mathematically in the modern computational sense; he did something earlier and equally revealing. He searched for the structure beneath the sacred text.

In *The Chronology of Ancient Kingdoms Amended*, published after his death in 1728, Newton reconstructed ancient history by comparing biblical records, classical sources, historical sequences, and astronomical calculations, convinced that if chronology was wrong, then the relationship between scripture, empire, prophecy, and providence would also be misunderstood. In *Observations upon the Prophecies of Daniel, and the Apocalypse of St. John*, published in 1733, he treated prophecy not as mystical chaos but as a symbolic system governed by rules: beasts, horns, kingdoms, times, numbers, and historical fulfillments had patterns to be decoded with care. Newton's Bible was not merely a devotional text lying outside reason. It was, for him, a structured universe of meaning. This book begins from that same ancient impulse, now carried into a new age: if the greatest scientific mind of the early modern world sought the hidden architecture of scripture with the tools of his time, what might be possible today with the tools of computation, systems theory, mathematical modeling, and artificial intelligence?

For most of human history, religion was not a weekend hobby; it was the operating system of civilization. Kings ruled "by the will of God," markets were embedded in moral law, families and empires alike understood themselves through revelation. The political, social, and economic veins of society all flowed through a religious heart. Then, over just a few centuries, the Renaissance and the scientific revolution rewired that heart. Mathematics, experiment, and technology became the new arbiters of truth. Nietzsche captured the psychological shock of this shift with his famous verdict: "God is dead. God remains dead. And we have killed him."[1] It was not just a statement about belief; it was a diagnosis that the old religious foundations of knowledge and morality had collapsed in the modern West.

In the process, science quietly assumed the role that religion once held. It offered a method for formulating and testing hypotheses, a language of numbers and equations that could predict and control the physical world. Religion, by contrast, continued to speak in stories, symbols, and legal-ethical prose. Scientists became the new priesthood: they could walk into any public debate and, armed with data and experiment, dismiss religious claims as myth or untestable opinion. From natural sciences to social sciences, "rigor" meant mathematical formalization and empirical verification. When pseudo-scientific ideologies such as eugenics were weaponized by the Nazis to justify genocide, religion often found itself sidelined or unable to mount a compelling, structurally articulated counter-argument. Later, the same pattern emerged in new guises: scientific and technical language used to rationalize mass violence, and now, AI-driven targeting systems that can choose when and where to strike. Both science and religion have been taken hostage by power, but only science seemed to have a language that the modern world was willing to call "objective."

Religion, meanwhile, was pushed into a corner. Because it spoke in textual and symbolic language, it was easily portrayed as subjective, mythic, or irrational. As science advanced and technology transformed everyday life, religion's public credibility eroded. Atheism surged not just because people became more rebellious, but because scientific explanations were simple, tangible, and demonstrably useful. Religions, unable to articulate their claims in the dominant mathematical-causal idiom, retreated into two main positions: a privatized cultural religiosity that made its peace with secular structures, or a reactive extremism that tried to reassert authority through force rather than argument. In many Christian and Muslim societies, what had once been a civilizational backbone fractured into either secularized tradition or militant counterculture. A force that had guided humanity for millennia seemed, within four centuries, to have been demoted to the margins while science took over as the "real" religion of modern life.

Contemporary interpreters have not been shy about drawing the conclusion. Nietzsche saw early that once the traditional God was declared unbelievable, a vacuum would open at the center of moral and metaphysical life. In our own time, Yuval Noah Harari has suggested that a new creed is already rising to fill that void: "The most interesting emerging religion is Dataism, which venerates neither gods nor man – it worships data."[2] In his narrative, AI and big data will soon understand us better than we understand ourselves, and traditional religions will be replaced or absorbed into algorithmic systems and information flows. In other words: the future belongs to Silicon Valley, not Sinai; to the server farm, not the sanctuary.

We believe this diagnosis is only half right. It is correct that a new regime of intelligence is emerging, one in which machines participate in perception, prediction, and decision-making. It is correct that, left to itself, this regime will treat human beings as data-processing nodes in larger optimization problems. But it is wrong – deeply, historically wrong – to assume that religion must simply fade away or be replaced by a "better" synthetic creed designed by engineers. The real reason religion appears defenseless is not that its content is weak, but that its language has never been translated into the formal idioms that govern modern systems: computation and mathematics. Science speaks in equations and models; religion has, until now, spoken in verses, sermons, and legal manuals. That asymmetry is what makes religion look subjective and science look objective.

Our work begins to change that asymmetry. After millennia in which scripture has lived almost exclusively in textual, narrative, and juristic form, we are giving it a second language: the language of formal models – variables, functions, graphs, and dynamic systems. Under the umbrella of Computational Religious Studies – a new field we have pioneered – the Computational Quran, and the Computational Bible, we treat revelation not as a loose collection of inspirational sayings but as a structured epistemic system: a body of knowledge about the soul, society, history, and destiny that can be architected, modeled, and analyzed. We do not replace tafsīr or theology; we formalize their structural output. Concepts like belief, hypocrisy, justice, mercy, covenant, sin, and salvation become variables with defined domains. Patterns of rise and fall, repentance and hardening, oppression and collapse become dynamic trajectories and feedback loops that can be studied and simulated. For the first time, religion begins to speak fluently in the same mathematical-computational idiom that runs our sciences, our policies, and our machines.

This does not mean that religion is reduced to equations or that mystery is abolished. It means that where revelation itself offers clear patterns and structures, we now have the tools to express those patterns in a form that AI systems, social simulations, and policy models can actually use. In this frame, religion is no longer a passive, pre-modern inheritance sitting uneasily beside "real" knowledge. It becomes a source of world models: explicit architectures of moral and spiritual causality that can guide how we build and constrain intelligent systems. Our current work focuses on Islam and Christianity, not because others are irrelevant, but because they offer rich, living traditions and scriptural corpora that have profoundly shaped global civilization. In time, the same methods can be extended to other revelations and religious philosophies, with each modeled on its own terms and then brought into comparative dialogue.

The implication is as bold as it is simple: AI will not eliminate religion. With the right formalizations, AI will become one of the primary arenas in which religion does its work. Instead of being merely "trained on religious text" as an undifferentiated corpus, AI systems can be designed to live inside revelation-aware models – to reason within Quranic and Biblical architectures of justice and mercy, to evaluate actions against formally expressed moral constraints, and to help human beings see the long-run trajectories of their choices through the lens of their own faiths.

In that sense, our claim is unapologetically ambitious: religion is not only back; it is destined to be a central partner in the age of shared intelligence between humans and machines. By translating scriptures into mathematical and computational form, we are not capitulating to science – we are inviting science and technology to meet revelation on equal footing.

What our work does is convert this historical opening into a concrete pathway. Quantum mechanics, complexity science, and AI have already shattered the illusion that reality is a closed, mechanistic machine fully captured by classical equations; they gesture toward layers of indeterminacy, information, and consciousness that our current models struggle to explain. But without a structured way to bring revelation into this conversation, those hints either collapse into vague spirituality or are colonized by new secular mythologies about data, simulation, or "the universe" as an impersonal force. By giving religion a rigorous mathematical and computational language, CRS prevents that drift. It allows concepts like soul, justice, accountability, mercy, and transcendence to enter the same formal space as wave functions, state spaces, and learning algorithms. In doing so, it offers science a partner that can help it interpret its own discoveries within a higher architecture of meaning – and offers religion a way to lead, not by rejecting modern knowledge, but by structuring it toward humane and transcendent ends.

We also wish to express our deep gratitude to our publishers, whose confidence and support made it possible to bring this work into the world in its present form. They have shown patience with a project that is both unconventional and ambitious, and have given us the space to experiment with a new intellectual language at the intersection of scripture, mathematics, and artificial intelligence. Their commitment to thoughtful, boundary-pushing scholarship has been essential to the birth of this book and, we hope, to the emergence of Computational Religious Studies as a recognized field of inquiry.

We stand on the shoulders of generations of scholars in theology, philosophy, science, and mathematics whose efforts prepared the ground for this work. Classical exegetes, jurists, and theologians mapped the semantic and doctrinal terrain of the scriptures. Modern scholars of religion and the humanities refined the tools of hermeneutics and critical analysis. Scientists and mathematicians – across fields as diverse as physics, systems theory, information theory, logic, and computer science – developed the formal languages and conceptual frameworks that we now adapt to the study of revelation. Every equation in this book and every model we propose is, in some way, indebted to the intellectual labor of these communities. If

CRS represents something new, it is only because so much has already been built by others.

We also wish to be transparent about the tools we have used in the creation of this work. Throughout the development of our models and the drafting of these chapters, we made extensive use of Large Language Models – primarily ChatGPT, and in some places Gemini – to assist with editing, refinement of prose, organization of arguments, and the analysis, development, or verification of mathematical expressions. These systems did not originate the central concepts or frameworks, but they served as powerful instruments for exploring alternatives, checking consistency, and sharpening exposition. The American Institute of Artificial Intelligence fully endorses and permits such knowledge-creation activities using LLMs, provided they are directed, supervised, and grounded in human judgment, domain expertise, and ethical responsibility. In a book that argues for a deep integration between religion and computation, it is fitting that computational tools themselves played a role in shaping the text you now hold.

HOW TO READ THIS BOOK?

A word, finally, about the shape of this larger undertaking and where the present volume sits within it. This book is the conceptual and methodological foundation, it establishes the field, the framework, and the language. The models themselves, worked out in full, live in a companion volume: *The Computational Quran for the AI Era: Selected Mathematical Models of Divine Guidance*. That introductory edition presents eighteen carefully selected surahs, each modeled through a disciplined framework that is respectful of Islamic tradition and designed for computational clarity. It is not numerology and not metaphor. It treats revelation as a system, ensuring its meaning remains intelligible and protected in a world increasingly shaped by machines. Readers who wish to see the architecture of this book rendered into complete, working mathematical form will find those eighteen models there. And this is only the beginning: the eighteen surahs represent a first, deliberate selection, and we are committed to releasing the complete mathematical translation of the Qur'an in full as the work matures. What you hold explains why scripture can be modeled and how; the companion volume shows what those models look like when fully built; and the complete translation, in time, will carry the method across the entire text.

HOW TO GET ACCESS TO COMPUTATIONAL QURAN?

A word, finally, about the shape of this larger undertaking and where the present volume sits within it. This book is the conceptual and methodological foundation, it establishes the field, the framework, and the language. The models themselves, worked out in full, live in a companion volume: *The Computational Quran for the AI Era: Selected Mathematical Models of Divine Guidance (Naqvi and Naqvi 2026).* That introductory edition presents eighteen carefully selected surahs, each modeled through a disciplined framework that is respectful of Islamic tradition and designed for computational clarity. It is not numerology and not metaphor. It treats revelation as a system, ensuring its meaning remains intelligible and protected in a world increasingly shaped by machines. Readers who wish to see the architecture of this book rendered into complete, working mathematical form will find those eighteen models there. And this is only the beginning: the eighteen surahs represent a first, deliberate selection, and we are committed to releasing the complete mathematical translation of the Qur'an in full as the work matures. What you hold explains why scripture can be modeled and how; the companion volume shows what those models look like when fully built; and the complete translation, in time, will carry the method across the entire text.

CHAPTER OVERVIEW

The fifteen chapters of this book move in a deliberate arc: from the founding of a discipline, through the building of its tools, into the living demonstration of those tools upon the text itself, and outward at last to the horizons where revelation, society, and machine intelligence meet. The reader who travels the whole path will arrive somewhere different from where they began, having watched scripture pass from sacred language into structured, computable architecture without ever ceasing to be sacred. Below we trace that path, chapter by chapter, so that the traveler knows both the terrain ahead and why each step matters.

The journey opens, in *Chapter 1, Computational Religious Studies: The Birth of a New Field,* with the founding act itself. Here the stakes are named without flinching: humanity has built artificial cognition powerful enough to reshape belief, identity, and institutions at the very moment traditional faith is receding across much of the world. The chapter argues that a perpetually present machine intelligence cannot remain neutral toward revelation; left to itself, it will flatten scripture into one more stream of negotiable opinion. The answer it proposes is audacious and precise, translate revelation into the one language both human and machine must obey, mathematics, and from that conviction a new discipline is born. This chapter is the threshold; everything after it is consequence.

Chapter 2, The Upcoming AI Is Cognitive, widens the lens to the machines themselves. It argues that today's generative systems, for all their fluency, remain pattern-recognizers, and that the next frontier belongs to cognitive machines built on world models — internal representations of states, causes, consequences, and possible futures. The chapter's quiet warning is that if scripture reaches these systems only as raw text, they will mistake surface association for understanding. Its deeper promise is that the Computational Quran can supply what such machines will require: explicit moral-causal structure they can reason over. The reader leaves this chapter understanding not only where AI is going, but why revelation must be ready when it arrives.

With the stakes and the technological horizon established, *Chapter 3, Scripture as System,* supplies the conceptual key that makes the entire project coherent. Drawing on systems theory, it shows that the Quran can be understood as a living, open, dynamic system, one that exchanges inputs and outputs with the world of its readers and exhibits the feedback, adaptation, and emergence characteristic of complex adaptive systems. This is the chapter where the reader first sees that treating scripture as a system is not a metaphor imposed from outside but a recognition of structure already present within.

Chapter 4, Introduction to the Computational Quran, grounds the ambition in the theology of the text itself. It returns to first principles, revelation as *nazool,* the transfer of divine knowledge to human cognition; the Quran as *hidayat,* guidance made accessible to all; *tafsir* as the disciplined human labor of decoding it. By rooting the computational project in the Quran's own account of how its guidance is received and understood, this chapter reassures the reader that what comes next extends the interpretive tradition rather than displacing it.

Now the toolmaking begins. *Chapter 5, The Five-Step Modeling Framework,* lays down the core pipeline of the discipline: from preserved text, to tafsīr-grounded meaning, to variables, to formal mathematical structure, and finally to integration within a scripture-wide atlas. This is the methodological spine of the book, the repeatable procedure that ensures every model remains anchored to the text and traceable back to it. A reader who understands this chapter understands how all the models in the book were made.

Chapter 6, Mathematical Techniques in Scripture Modeling, hands the reader the instruments themselves, variables and functions, calculus, thresholds, vectors, graphs, probability, information theory, and logic, but always in service of intuition rather than intimidation. Its governing insight is that the mathematics is not imposed upon scripture; it emerges from the relationships the text already describes, between belief and action, cause and consequence, individual choice and collective fate. Mathematics here is not a replacement for revelation but a second language for its structure.

Chapter 7, The Universal Quranic Variables, supplies the vocabulary that flows through the pipeline. It introduces the Universal Quranic Variable Ontology, a standardized, evolving taxonomy of core Quranic concepts rendered as consistent, machine-readable variables. If the framework is the grammar and the techniques are the syntax, this ontology is the lexicon: the shared semantic backbone that lets thousands of verse-level models speak to one another without fragmenting into incoherence. It is among the most quietly consequential chapters in the book.

Then comes the crossing. *Chapter 8, Modeling Surah al-Fāti ah: The Opening Function*, carries the reader from theory *about* the method into the practice *of* it. Working through al-Fāti ah and the opening verses of al-Baqarah, it demonstrates how verses the reader already knows by heart can be expressed as structured systems of variables and equations without losing their spiritual power — guidance as a vector, hypocrisy as an oscillation, repentance as an impulse in a moral control system. This is the chapter where the abstract becomes vivid, and the promise of the earlier chapters is finally seen at work.

Chapter 9, Algorithmic Reasoning, AI Agents, and Simulations in Computational Quran, sets the models in motion. It shows the Computational Quran not merely admired but used: a question-answering agent assisting a finance ethics board, an alignment checker evaluating an AI system against Quranic criteria for just speech, simulations that trace ethical dynamics over time. Crucially, it frames the Computational Quran as a durable model layer, one that can sit beneath today's language models and tomorrow's world models and quantum algorithms alike. Here the reader sees that the structures built in earlier chapters are not static diagrams but working machinery.

Chapter 10, Applied Quran: Reforming the Social Sciences, marks the project's most expansive turn. It argues that the Quran can serve as a foundational framework for social science itself, supplying axioms, social objectives, and testable causal propositions through which poverty, governance, social cohesion, and civilizational resilience can be studied. Two worked examples, built on open-source intelligence and the framework's own harm taxonomy, demonstrate the method on real terrain: a cross-national analysis of institutional hypocrisy and a systems-level account of social decline under deception and fragmentation. The chapter's boldest claim is that all social science begins with assumptions, and a Quranic framework simply makes its own explicit and tests what follows.

Chapter 11, From Scripture to Machine: Integrating the Computational Quran with AI, returns to the confrontation that opened the book and resolves it constructively. As intelligence is increasingly externalized into machines, it argues, those machines must have structured access to what it means to be human, and religion is too central a dimension of human life to be excluded or fed in naively. The Computational Quran offers a third path between exclusion and distortion: an intermediate language through which revelation becomes computationally accessible without forfeiting its theological integrity. This is the chapter that shows how the discipline's tools actually enter the machine.

The final movement turns toward implications. *Chapter 12, CRS and Interpreting Scriptures*, confronts the oldest problem of all. that the same sacred words have been read to justify both liberation and domination, mercy and cruelty. It proposes that the discipline can convert interpretive claims into reproducible, transparent, analyzable models, giving interpretation a shared structural ground on which positions can be compared and examined rather than merely asserted. It does not replace classical interpretation; it makes its hidden architecture visible.

Chapter 13, CRS and Comparative Religion, extends that structural clarity across traditions. Where comparative religion has long relied on blunt abstractions and the quiet projection of one tradition's categories onto another, this chapter offers a variable-based lens: each tradition modeled first on its own terms, then aligned and compared at the level of variables and dynamics. The aim is not to erase real theological difference but to make it intelligible — to compare structures rather than impressions.

Chapter 14, The Future of Computational Quran: Complex Adaptive Systems and Quantum Computing, lifts the reader's gaze to the frontier. Having become a concrete object, an interconnected system of variables, equations, and semantic structures, the Computational Quran is now placed inside the two revolutions reshaping our understanding of complexity and computation. The chapter sketches how Quranic models might serve as rule sets in large-scale simulations of moral and social dynamics, and how quantum representations might illuminate the vast, high-dimensional space of Quranic knowledge graphs, all while holding fast to the theological boundaries drawn earlier.

The journey closes where it must, with *Chapter 15, CRS: The Science of Revelation*. It reframes the provocative question with which the whole enterprise began — can revelation be a science? It answers carefully: not a secular science that explains religion away as mere human phenomenon, but a science *of* revelation that treats the revealed text as an authoritative specification of how reality is and ought to be, and asks how that specification can be expressed in rigorous, formal terms. It is the chapter that gathers the entire book into a single conviction: that revealed architectures can, and must, enter the computational age on their own terms, leading rather than being led.

Read in sequence, these fifteen chapters form one continuous argument, from the birth of a field to the science it makes possible. But no door here is locked. The theologian may enter through the chapters on interpretation and comparison; the engineer through the framework, the techniques, and the integration with AI; the curious believer simply at the beginning, walking the path as it unfolds. However one enters, the destination is the same, a way of seeing scripture, for the first time, by a new kind of light.

REFERENCES

Naqvi, A., & Naqvi, N. (2026). *The Computational Quran for the AI era: Selected Mathematical Models of Divine Guidance*. American Institute for AI & Quantum.

ENDNOTES

1 https://philosophybreak.com/articles/god-is-dead-nietzsche-famous-statement-explained/

2 https://www.goodreads.com/quotes/8838687-the-most-interesting-emerging-religion-is-dataism-which-venerates-neither

Acknowledgement

A work of this kind is never the product of its authors alone. It rests on the patience, skill, and goodwill of many hands, and it is our privilege to acknowledge them here.

We owe a special debt of gratitude to **Marya Naqvi**, whose contribution to the Computational Quran work was both substantial and indispensable. The clarity and discipline with which the formal material is presented owe a great deal to her support and diligence. For her steady hand and her commitment to seeing this work properly into the world, we are deeply thankful.

We are likewise grateful to the reviewers whose careful reading strengthened this book. Their thoughtful questions, corrections, and challenges sharpened the arguments and improved the flow and structure of the work. A project that proposes a new field invites scrutiny, and we are fortunate to have received scrutiny of such quality. Whatever rigor these pages possess is in no small part owed to their engagement.

We also wish to thank our publisher, **IGI Global Scientific Publishing**, for its confidence in a project that is at once unconventional and ambitious, and for the space it gave us to develop a new intellectual language at the intersection of scripture, mathematics, and artificial intelligence.

We are grateful, finally, to the scholars of *tafsīr* and the broader traditions of theology, mathematics, and computer science upon whose labor this work quietly depends. Whatever is sound in these pages we owe in part to them; whatever falls short is our own.

Section 1
Foundations of a New Discipline: Computational Religious Studies

Chapter 1

Computational Religious Studies (CRS): The Birth of a New Field

ABSTRACT

This chapter introduces Computational Religious Studies (CRS) as a new discipline designed to safeguard and empower religion in the age of artificial intelligence. As AI threatens to flatten revelation into mere data, CRS translates scripture into mathematical and computational structures that preserve its semantic integrity while making it intelligible to machines. The chapter contrasts CRS with traditional theology and religious studies, outlines its methodological foundations, and situates it within the long history of attempts to study religion. By establishing principles, tools, and modeling frameworks, the chapter presents CRS as a transformative paradigm that enables faith to operate as a rigorous, world-shaping epistemic system in modern scientific and technological contexts.

INTRODUCTION

Humanity now stands at the edge of its greatest trial: we have created artificial cognition powerful enough to shape our choices, identities, and institutions at the very moment religious affiliation is collapsing across much of the world (Krügel *et al.* 2023, Faber *et al.* 2024). The internet has already weakened traditional belief and community by flooding individuals with endless alternatives, undermining religious authority, and normalizing life without God. AI will go further (Downey 2014, McClure 2020, Campbell 2007). A constantly present cognitive engine – consulted for work, guidance, and even moral judgment – cannot remain neutral with respect to faith. It will ingest scripture as just another text stream, reduce revelation to patterns of words, and place divine speech on the same plane as opinion,

DOI: 10.4018/979-8-3373-6940-2.ch001

ideology, and entertainment. It will "understand" all doctrines only as competing narratives, value them all as equally negotiable, and in practice privilege secular and technocratic frames over revealed truth.

If religion remains only in natural language, it will be at the mercy of these artificial minds. The only way to protect revelation in an age of machine cognition is to translate and express it in the one language both human and machine must obey: mathematics. That is the foundational purpose of this project and the field we are founding – Computational Religious Studies (CRS). By encoding scripture as rigorous mathematical structures – systems of variables, functions, constraints, and dynamics – we achieve two non-negotiable outcomes. First, the core logic of religion is forever shielded from distortion: AI systems may process texts however they wish, but the canonical mathematical models stand as a precise, testable, and incorruptible reference of doctrinal meaning. Second, these same structures become directly usable by AI itself, enabling the construction of divinely guided artificial cognition – systems whose reasoning is "divine-native," not merely scripture-flavored. CRS is therefore not an academic curiosity; it is a strategic defense of faith in the age of artificial minds and the blueprint for ensuring that future intelligences can reason *with* revelation rather than silently erasing it.

We are launching a new field: Computational Religious Studies (CRS). As pioneers of this discipline, our aim is not simply to help religion survive in the AI era, but to position it as the central architecture around which artificial cognition is shaped and human–machine collaboration unfolds. To make that possible, we have undertaken something unprecedented: translating sacred scriptures into the language of mathematics. Our work currently spans both the Qur'an and the Bible. This volume focuses on the Computational Quran; the next will present the Computational Bible.

Wherever knowledge takes us – however it is represented and in whatever form cognition manifests, from graphs to quantum substrates and beyond – we can be certain that religion and the divine intention for human life and social order will remain integrated and alive. With CRS, artificial cognition will not merely talk *about* God; it will reason *through* the structures of revelation itself, so that machine intelligence can, at last, speak with a rigorously grounded divine voice.

SCIENCE VS. RELIGION

One of the most unfortunate consequences of the scientific revolution was the formal partition between religion and science – a separation that set them on divergent paths. Science, in its pursuit of empirical discovery, advanced human capability through technological and natural sciences, yielding cures for diseases, industrial revolutions, and remarkable feats of engineering. Yet, stripped of the moral and civilizational compass that religion offers, these same advances also paved the way for the most destructive episodes in human history: two world wars, countless regional conflicts, the atomic bomb, and the accelerating crises of climate change and environmental degradation. Today, credible voices warn that humanity is closer to nuclear annihilation than ever before, and that unchecked ecological collapse may render the Earth inhospitable for future generations.

This predicament is not solely the fault of science – it is equally attributable to the inability, and in many cases unwillingness, of religious intellectuals to engage in epistemic innovation. The Qur'an repeatedly calls upon believers to reflect, inquire, and pursue knowledge (e.g., "Do they not reflect upon themselves?" 30:8; "Do they not contemplate the Qur'an?" 4:82), and the Bible issues similar commands to active, reflective engagement with revelation ("Reflect on what I am saying, for the Lord will give you insight into all this," 2 Timothy 2:7; "but test everything; hold fast what is good," 1 Thessalonians 5:21). Yet for centuries, the intellectual apparatus of religion remained stagnant. Religious scholarship largely confined itself to preservation and commentary, while avoiding the development of new frameworks capable of engaging with – and shaping – the trajectory of human progress. This absence of structured inquiry and methodological renewal created a paralysis within religious thought, allowing secular ideologies to dominate the architecture of human systems.

Even as we write this book, we recognize the magnitude of the challenge before us. For many, the notion that religion can undergo epistemic innovation through computational methods will be unfamiliar, even unsettling. Yet this is precisely the work that must be undertaken if religion is to reclaim its rightful role as a positive, constructive force in guiding human civilization – one that does not compete with science but stands as its ethical and conceptual equal, informing it, humanizing it, and restraining its destructive potential.

Computational Religious Studies (CRS) emerges precisely to address this longstanding rupture. It does not seek to "modernize" religion in the sense of altering its core truths, nor to subordinate it to scientific paradigms. Rather, it creates a third epistemic framework – one that enables religion to engage with science on equal footing through formal modeling, systems theory, and computational logic. In the Qur'an, we find repeated references to the harmony between divine revelation and the created order: "We will show them Our signs in the horizons and within themselves until it becomes clear to them that it is the truth" (41:53). A parallel vision is expressed in the Bible: "The heavens declare the glory of God; the skies proclaim the work of his hands" (Psalm 19:1), and "For since the creation of the world God's invisible qualities… have been clearly seen, being understood from what has been made" (Romans 1:20). Together, these texts invite a synthesis in which the "signs" discovered through empirical inquiry are interpreted and integrated within the moral and existential framework of revelation. CRS operationalizes this synthesis by providing religion with an updated methodological toolkit – one that does not merely digitize sacred texts, but models their conceptual structures, maps their variables, and renders them intelligible to both human scholars and artificial intelligence systems. In doing so, CRS transforms religion from a passive observer in the modern world into an active, adaptive participant in shaping humanity's future.

THE INTELLECTUAL FOUNDATIONS OF RELIGION

Religion, in all its richness and complexity, has been defined and understood in countless ways throughout history. The diversity of definitions reflects the impossibility of capturing its totality in a single formulation. Broadly speaking, it encompasses a collection of cultural systems, belief systems, and worldviews that connect humanity to spirituality and, often, to moral values. More than a single institution or dogma, religion can be seen as a multidimensional phenomenon comprising interrelated elements:

Beliefs. At its core, religion often involves belief in the supernatural, the transcendental, or the spiritual – whether God, gods, spirits, or a divine power. These beliefs form the epistemic backbone of religious life, offering frameworks for understanding existence, purpose, morality, and the mysteries of life and death. The Qur'an captures this dimension in the declaration: "Indeed, this Qur'an guides to that which is most upright" (17:9), highlighting belief as a compass for navigating reality. The Bible expresses a parallel vision of belief as both orientation and way of knowing: "Now faith is the assurance of things hoped for, the conviction of things not seen" (Hebrews 11:1), and "Your word is a lamp to my feet and a light to my path" (Psalm 119:105). Together, these scriptural witnesses present belief not as blind assent, but as a structured way of seeing and walking through reality.

Practices and Rituals. Religions are also defined by structured behaviors – prayer, fasting, meditation, worship, pilgrimage – that serve to embody belief in action. These rituals are not merely symbolic; they reinforce cognitive and moral commitments, bind communities, and translate abstract values into lived experience. The Qur'an prescribes such acts not as ends in themselves, but as means of cultivating taqwā (God-consciousness), as in fasting: "...that you may become mindful of God" (2:183). The Bible likewise frames ritual practice as formative rather than mechanical: Jesus teaches, "When you pray..." (Matthew 6:5) and "When you fast..." (Matthew 6:16), assuming these disciplines as integral to faithful life, while Paul exhorts believers to "present your bodies as a living sacrifice, holy and acceptable to God – this is your spiritual worship" (Romans 12:1). In both traditions, embodied practices function as structured pathways through which belief shapes identity, character, and community.

Community. A religious tradition naturally fosters a community of adherents bound by shared convictions and practices. This social dimension provides belonging, identity, and mutual support. The Qur'an addresses believers collectively – "Indeed, the believers are but brothers" (49:10) – signifying that religious life is not meant to be solitary, but embedded in a network of mutual responsibility. The Bible offers a parallel vision of communal identity: "Now you are the body of Christ, and individually members of it" (1 Corinthians 12:27), and it portrays the early believers as those who "devoted themselves to the apostles' teaching and to fellowship, to the breaking of bread and to prayer" (Acts 2:42). In both traditions, faith crystallizes not only as personal conviction, but as a shared life woven through bonds of loyalty, care, and collective worship.

Morality and Ethics. Many religions provide ethical systems that govern personal conduct and social relations, often rooted in divine command or sacred law. These moral codes are not arbitrary but claim authority from revelation, prophecy, or divine wisdom. In the Qur'an, ethics are inseparable from belief: "Indeed, Allah commands justice, excellence, and giving to relatives, and forbids immorality, bad conduct, and oppression" (16:90). The Bible articulates a similarly integrated moral vision: "He has shown you, O man, what is good... to act justly and to love mercy and to walk humbly with your God" (Micah 6:8), and Jesus summarizes the law and the prophets in the twin commands to love God and neighbor (Matthew 22:37–40). In both scriptures, ethics are not an optional add-on to faith, but the concrete expression of what it means to live in right relationship with God and with others.

Sacred Texts and Narratives. Every religious tradition contains sacred narratives – textual or oral – that explain origins, articulate moral truths, and guide behavior. These are not static records; they are dynamic repositories of meaning, passed across generations through recitation, teaching, and storytelling. In Islam, the Qur'an is both text and living discourse, meant to be recited, pondered, and applied: "Then do they not reflect upon the Qur'an...?" (47:24). The Bible likewise presents scripture as a living word that teaches, shapes, and corrects: "All Scripture is breathed out by God and profitable for teaching, for reproof, for correction, and for training in righteousness" (2 Timothy 3:16), and "For everything that was written in the past was written to teach us..." (Romans 15:4). In both traditions, sacred texts function as ongoing, performative narratives through which communities continually renegotiate their understanding of God, self, and world.

Symbols and Places. Religion finds expression through symbols, art, music, and architecture, and often designates specific places as sacred. These serve as tangible focal points for identity, devotion, and cultural memory. The Qur'an affirms the importance of such sacred markers in the notion of *sha'ā ir Allāh* (the symbols of God): "That, and whoever honors the symbols of Allah – indeed, it is from the piety of hearts" (22:32). The Bible likewise institutes symbolic acts and sacred markers as enduring signs, as when Joshua sets up memorial stones at the Jordan "so that this may be a sign among you" (Joshua 4:6–7). In both traditions, symbols and sacred places function as visible anchors of an invisible allegiance, binding communities to a shared spiritual history.

Social, Political, Economic System: Religion also operates as a comprehensive social, political, psychological, legal, and economic system that can help organize and guide entire societies. Far from being confined to private belief or occasional ritual, it offers architectures for governance, dispute resolution, welfare, and collective responsibility. The Qur'an repeatedly frames divine guidance in institutional and structural terms: commanding the just distribution of trusts and judgments (4:58), insisting on public justice even against one's own interests (4:135), urging consultative governance among believers (42:38), prohibiting economic exploitation and corruption in transactions (2:188; 2:275–279), and mandating mechanisms like zakat and charity to sustain social welfare. The Bible, likewise, presents faith as a public order with legal and economic implications: it envisions rulers as "servants of God for your good" tasked with upholding justice (Romans 13:1–4), portrays righteousness and justice as forces that exalt and stabilize a nation (Proverbs 14:34; Amos 5:24), and depicts early believers sharing resources so that "there was not a needy person among them" (Acts 4:32–35). In both traditions, religion is not merely a private spirituality but a blueprint for shaping laws, institutions, economic flows, and psychological well-being – aiming to align the inner life of individuals with the outer structures of society.

From a historical perspective, the very concept of "religion" as a discrete category is relatively modern, emerging largely from Western intellectual traditions. In pre-modern contexts, religion was fully integrated into all aspects of life – politics, law, science, economics – making no sharp distinctions between sacred and secular. The Latin root *religiō* has been interpreted as "to bind together" or "to read again," both implying a duty of continual engagement and alignment with the divine order.

From a sociological standpoint, religion functions as a potent social force: promoting cohesion, reinforcing shared values, and offering purpose and meaning. It can also serve as a stabilizing agent, encouraging moral behavior and upholding social order. Yet, in the absence of adaptive intellectual renewal, religion can risk becoming a static repository of tradition rather than an active architect of societal direction.

A BRIEF HISTORY OF THE STUDY OF RELIGION

The study of religion has undergone a remarkable transformation – from a primarily theological and philosophical pursuit to a formal, interdisciplinary academic field. Across history, the lens through which religion has been examined has shifted repeatedly, reflecting changes in intellectual climate, methodological innovation, and broader cultural currents.

Early and Pre-Modern Approaches

In the ancient and classical worlds, early thinkers such as Herodotus and Hecataeus of Miletus observed and described the religious practices of other cultures (Özcan, 2019). Their work was primarily descriptive, seeking to catalog beliefs, rituals, and deities, often with the implicit aim of understanding the "other" rather than reformulating their own religious frameworks.

During the medieval and Renaissance periods, the study of religion remained largely confined to theology within religious institutions. Islamic and Christian scholars alike devoted themselves to the interpretation of their own sacred texts and traditions, but there were notable figures – such as al-Shahrastānī (1923), who documented the creeds and sects of various peoples, and Peter the Venerable, who commissioned Latin translations of Islamic texts – who engaged with other religions, albeit often with apologetic or polemical intent (Kritzeck, 1964).

THE RISE OF THE "SCIENCE OF RELIGION" IN THE 19TH CENTURY

The 19th century marked a decisive shift, as the academic study of religion emerged as a field distinct from theology. This "scientific" approach sought to apply the methods of other newly professionalized disciplines to the study of religion, treating it as a human phenomenon subject to systematic analysis (Byrne, n.d.; Sharpe, 1986).

- *Comparative Philology and Mythology.* Max Müller, often called a founder of the discipline, used comparative philology to examine religious texts across cultures (Müller, 1873; Müller, 1878). He argued that many religious ideas arose from the mythic interpretation of natural phenomena, tracing their linguistic roots and symbolic parallels.
- *Evolutionary Anthropology.* Scholars such as Edward Burnett Tylor and James Frazer advanced evolutionary models of religion, suggesting that it developed through stages – animism, polytheism, and eventually monotheism – each reflecting changes in human cognition and social organization (Tylor, 1871; Frazer 1890).

- *Sociology of Religion.* Émile Durkheim reconceived religion as a fundamentally social phenomenon, proposing that rituals and beliefs served to reinforce collective bonds and a shared moral framework, with the ultimate object of worship being society itself (Durkheim, 1995).
- *Psychological Interpretations.* Sigmund Freud and others approached religion as a psychological construct, interpreting it as an expression of subconscious needs, fears, and desires – an "illusion" that provided comfort and a sense of order in the face of existential uncertainty (Freud, 1961; Freud, 1913).

20th Century and Contemporary Developments

The 20th century saw a shift toward more nuanced, context-sensitive approaches.

- *Phenomenology of Religion.* Scholars like Rudolf Otto and Mircea Eliade emphasized the lived experience of religion, seeking to understand it from the perspective of practitioners. Otto's concept of the *numinous* – a unique feeling of awe and mystery – became central to this approach, as did Eliade's focus on the sacred as a distinct category of human experience (Otto, 1923; Eliade, 1959).
- *Anthropological Fieldwork.* Anthropologists adopted "participant observation" as a core method, living among the communities they studied to gain deeper insight into how religion was practiced and how it intertwined with politics, economics, kinship, and daily life (Evans-Pritchard, 1937).
- *Post-Colonial and Critical Theories.* In more recent decades, post-colonial scholarship has questioned the very categories used to study religion, noting that the term "religion" itself is rooted in Western intellectual history and has often served as a tool for classifying – and subordinating – non-Western belief systems (Asad, 1993).
- *Interdisciplinary and Methodological Pluralism.* Today, the academic study of religion draws on a wide range of disciplines – history, anthropology, sociology, psychology, philosophy, literary criticism, art history, and increasingly, cognitive science. Scholars use an equally diverse set of methods, from textual criticism and archaeology to ethnography and computational modeling, to examine the complex and evolving phenomenon of religion (Pyysiäinen, 2013; Lane & Shults, 2021).

Across these developments, the field has grown more self-aware and methodologically plural, yet one limitation has persisted: even with advanced tools, the study of religion has rarely sought to structurally model its epistemic systems in a way that is both faithful to the internal logic of a tradition and translatable into formal, computable structures. It is here that Computational Religious Studies (CRS) positions itself – not as another lens in the history of religious studies, but as the next paradigm shift, one that can integrate theology, history, and lived experience into a coherent, operational framework for both human and machine reasoning.

Computational Religious Studies (CRS) differs from the computational science of religion (CSR) at its foundation: CRS starts *inside* the scripture as its primary object, while CSR starts *outside* religion as a human phenomenon. CSR treats religion as data – beliefs, rituals, networks – and uses cognitive theory plus computational tools (agent-based models, neural nets, game theory, text mining) to explain why and how religious behavior emerges, spreads, or changes. Its core question is descriptive and explanatory: "How do religious phenomena arise from human cognition and social dynamics?" CRS, by contrast, treats the Qur'an (and later, other scriptures) as a formal source of axioms, variables, and equations and systematically translates its guidance, narratives, and laws into explicit mathematical structures – functions, dynamic systems, graphs, and logical schemas. The primary question of CRS is not "Why do humans become religious?" but "What *computational system* of ethics, law, psychology, social order, and metaphysics does the scripture itself encode, and how can that system be made machine-readable and operational?" CSR uses computation to study religion as an object of science; CRS uses mathematics and computation to express religion as a scientific, model-bearing system in its own right, suitable for direct use in AI, policy design, simulations, and decision-support grounded in scripture.

THEOLOGY AND RELIGIOUS STUDIES

Religious Studies examines religion as a human phenomenon through secular, academic methodologies such as history, sociology, anthropology, psychology, and comparative literature. Its aim is to describe, compare, interpret, and explain religious beliefs, practices, institutions, and experiences without presupposing the truth claims of any particular tradition. The perspective is deliberately "from the outside." A scholar in this field might analyze the role of a ritual in maintaining social cohesion, or trace the historical compilation of a sacred text and its textual variants over time. The focus is on explaining religion's form, function, and evolution within the broader fabric of human culture (Morreall & Sonn, 2011; Orsi, 2003; Smart, n.d.).

Theology, by contrast, operates "from the inside." It uses the framework of a specific religious tradition to analyze, interpret, and systematize its teachings, doctrines, and sacred texts. Theology is concerned not only with understanding but with affirming a tradition's truth claims, providing guidance for believers, and answering questions such as: "What is God's will for humanity?" or "How should we live in accordance with divine revelation?" This requires an acceptance – at least provisionally – of the metaphysical claims that the faith puts forward (McGrath, 2017a; Chryssides and Geaves, 2007)

Put simply:

- Religious Studies asks, "What is the social function of this ritual?" or "How did this sacred text emerge, and why has it changed over time?"
- Theology asks, "What does this ritual mean for the believer's relationship with God?" or "How should we live according to this revealed text?"

While theology can be examined as an object of study within religious studies (for example, a historian might analyze the development of Christian theology), the two disciplines differ fundamentally in their starting points, assumptions, and aims. One seeks explanation without commitment; the other seeks understanding through commitment.

This divide is significant because neither framework, on its own, fully addresses the Qur'anic injunction to *"reflect"* (*yatadabbarūn*, 4:82) and *"ponder deeply"* (*yatafakkarūn*, 16:44) upon both revelation and creation. Religious Studies may offer analytic tools without embracing revelation's internal logic, while Theology may uphold revelation's truth but resist new epistemic methods. Computational Religious Studies (CRS) emerges as a third position – one that respects the internal coherence of a tradition while equipping it with formal, computational models that enable rigorous analysis, simulation, and application in dialogue with the sciences.

How Theology Has Been Studied

The study of theology has a deep and varied history, closely intertwined with the evolution of specific religious traditions and their institutional frameworks. At its core, theology has always sought to articulate, defend, and systematize a tradition's understanding of God, humanity, and the world. Yet the methods, emphases, and scope of theological inquiry have shifted dramatically across the centuries.

Ancient and Patristic Periods (Early Christianity). Early Christian thinkers – known as the Church Fathers – such as Augustine of Hippo and Athanasius of Alexandria, developed some of the first systematic theological framework (McGrath, 2017b). Their methods revolved around three primary activities:

- **Biblical exegesis**: interpreting sacred texts to extract doctrinal meaning.
- **Apologetics**: defending the faith against critics, both internal (heretical movements) and external (pagan philosophers, competing religions).

- **Doctrinal formulation**: articulating creeds and defining orthodoxy in response to theological disputes.

In this period, theology was inseparable from the life of the Church; it was not an abstract discipline but a living practice of teaching, defending, and guiding the faithful.

Medieval Scholasticism. The medieval period marked a turning point with the rise of scholasticism. Theologians such as Thomas Aquinas, Anselm of Canterbury, and Duns Scotus integrated classical philosophy – particularly Aristotle – into theological reasoning. Scholasticism elevated theology to what was then considered a *science* (*scientia* in the pre-modern sense): a systematic, rational body of knowledge grounded in first principles (in this case, divine revelation) and logically deduced conclusions. The scholastic method was methodical:

1. Present a precise theological question (e.g., *Does God exist?*).
2. Cite objections from authoritative sources, both sympathetic and critical.
3. Offer a counter-argument grounded in Scripture, tradition, or reason.
4. Develop a detailed, logical resolution to the question.
5. Respond explicitly to the original objections.

This formalized approach reflected a belief that theology, though grounded in revelation, could be organized and defended with the same intellectual rigor as other sciences of the time.

The Reformation and Post-Reformation. The Protestant Reformation reoriented theology around the principle of *Sola Scriptura* – Scripture alone as the ultimate authority. Reformers like Martin Luther and John Calvin emphasized direct engagement with biblical texts, often rejecting scholastic excess in favor of clarity, accessibility, and pastoral application. This era produced new systematic theologies that defined the doctrinal boundaries of emerging Protestant denominations, often in polemical contrast to Catholic teaching.

Modern and Contemporary Theology. In the modern era, theology has expanded to engage with a vast range of intellectual, scientific, and social developments:

- **Philosophical Movements**: Responding to Enlightenment rationalism, German idealism, existentialism, and postmodern critiques.
- **Scientific Discoveries**: Grappling with evolutionary biology, cosmology, and new conceptions of the universe.
- **Social and Political Issues**: Developing contextual theologies such as liberation theology, feminist theology, and post-colonial theology that address justice, inequality, and human rights.

- **Ecumenical and Interfaith Dialogue**: Opening conversations across denominational and religious boundaries to foster mutual understanding and theological exchange.

The "Science of Theology." The phrase "science of theology" historically referred not to laboratory experimentation, but to theology's claim to be a coherent, organized discipline capable of producing knowledge from first principles through logical reasoning (Kerr, 2002). In medieval thought, a *science* was any systematic body of knowledge grounded in certain foundational truths – axioms in mathematics, causes in natural philosophy, or divine revelation in theology. As Aquinas put it, theology is the *queen of the sciences* because its first principles are not derived from human reason but from God's own self-disclosure. While modern secular academia no longer places theology in this position of primacy, the term remains in use within religious traditions to describe theology's methodological rigor and its aspiration to order and clarity.

This historical trajectory shows that theology has continually adapted its tools and emphases in response to cultural and intellectual change – yet it has always remained tied to the internal logic of a specific tradition. Computational Religious Studies does not replace theology, but it offers it a new methodological partner: a way to preserve theological integrity while equipping it with formal, computational frameworks capable of mapping, modeling, and applying doctrinal systems in dialogue with modern sciences.

The "Science of Theology"

The term "science of theology" has a long history, and its meaning has shifted significantly over time, reflecting broader changes in how both "science" and "theology" are understood.

Medieval Meaning. For thinkers such as Thomas Aquinas, *scientia* (science) referred not to laboratory experimentation or empirical testing, but to a systematically organized body of knowledge derived from self-evident first principles (Kerr, 2002). In Aquinas's framework, theology was the highest of the sciences because its first principles came from divine revelation – truths given by God, who cannot err. These truths were considered more certain than anything derived from human reason or sensory observation. The task of the theologian was to use reason to interpret, clarify, and systematically draw conclusions from these revealed truths. In this sense, theology's "science" was deductive: beginning from divine premises and reasoning carefully to coherent conclusions.

Modern Meaning. Today, the phrase "science of theology" is used in more varied and contested ways (McGrath, 2001-2003):

- **A "Scientific" Approach to Method.** Contemporary theologians such as Alister McGrath use the term "scientific theology" to describe a methodological discipline. This approach does not claim that theology can be tested like physics or chemistry. Instead, it emphasizes clarity in articulating assumptions, transparency in methods, and rigor in evaluating theological claims against coherent criteria. It treats theology as a disciplined, critical, and logically structured enterprise, even if its starting points rest on revelation rather than empirical observation.
- **The Conflict with Modern Science.** Other scholars and critics contend that theology cannot be considered a "science" in the modern sense at all, since it is grounded in faith, revelation, and tradition rather than empirical evidence and falsifiable hypotheses. From this perspective, theology and the natural sciences are fundamentally different domains, each with distinct rules, goals, and forms of validation (Ford, 2005).

While both theology and religious studies require intellectual rigor, their starting points and ultimate aims are distinct. Religious studies begins from an external, secular standpoint, seeking to explain religion as a human phenomenon; theology begins within the framework of a tradition, seeking to articulate and apply that tradition's truths to life and society. The study of theology remains an ongoing, dynamic process – interpreting, systematizing, and applying a faith's core convictions in changing historical and cultural contexts.

From the perspective of Computational Religious Studies, this distinction is crucial. CRS does not attempt to make theology conform to the empirical sciences, nor does it dilute theology's faith commitments. Rather, it treats theological systems as structured bodies of knowledge that can be formally modeled, mapped, and analyzed – preserving their internal logic while enabling them to interact with other knowledge domains, including the sciences, through computational frameworks.

EXTENDING THE "SCIENCE OF THEOLOGY" INTO THE COMPUTATIONAL AGE

Computational Religious Studies (CRS) offers a way to bring the "science of theology" into the 21st century – not by altering its doctrinal truths, but by equipping it with formal, computational tools that make its reasoning structures explicit, analyzable, and applicable in new contexts. Historically, theology as *scientia* organized divine truths into coherent, logical systems, deducing conclusions from revealed first principles. CRS builds upon this legacy by treating theology itself as a structured knowledge system – one that can be mapped, modeled, and even simulated while remaining faithful to its revelatory foundation.

In practical terms, CRS approaches a theological tradition much as a systems theorist would approach a complex, rule-governed domain. The first step is **knowledge structuring**: identifying the axioms (revealed truths), the derived propositions (doctrinal conclusions), and the logical relationships between them. These elements can be expressed in formal, machine-readable models that preserve the internal logic of the tradition. For Islamic theology, for example, this could involve computationally representing doctrines of divine attributes, human responsibility, and eschatology in ways that mirror the reasoning of *kalam* while making them navigable by artificial intelligence systems.

Once these models are formalized, they can be simulated to explore theological reasoning under different conditions or in dialogue with new knowledge domains. For instance, a computational model could examine how a classical theological framework addresses contemporary ethical dilemmas or scientific findings, tracing the reasoning pathways to identify points of harmony or tension. Importantly, this simulation does not replace the theologian's interpretive authority; rather, it provides a tool for deeper analysis, enabling scholars and faith practitioners to see how their own doctrinal systems operate in a variety of intellectual environments.

CRS also allows for comparative computational theology, in which different traditions' reasoning structures can be mapped side-by-side – not to homogenize them, but to clarify their distinct epistemic architectures. Such comparisons can highlight where traditions share analogous logical forms (e.g., reconciling divine sovereignty and human agency) and where they diverge irreducibly.

By operationalizing theological reasoning in this way, CRS honors the medieval vision of theology as a science grounded in first principles while extending its methodological reach. It retains the faith commitments of a tradition by preserving its starting axioms, yet it enables rigorous analysis, testing, and application through computational means. In this sense, CRS is not an external imposition on theology but the next stage in its intellectual maturation – a modern set of instruments for an ancient and enduring science.

FOUNDATIONAL PRINCIPLES AND BEST PRACTICES FOR COMPUTATIONAL RELIGIOUS STUDIES (CRS)

As the founders of Computational Religious Studies, it is critical to establish a set of guiding principles that preserve the integrity of the field, ensure methodological rigor, and foster a respectful yet innovative approach to religious knowledge. The following ten ground rules define the ethos, scope, and practices of CRS.

1. Maintain Unwavering Respect for Religion and Sacred Texts.

At all stages of research, CRS practitioners must treat religious traditions and their scriptures with intellectual and moral respect. The work of CRS is not to diminish or trivialize sacred texts, but to approach them with the seriousness, reverence, and care they deserve as foundational sources of human civilization.

2. Focus on Applying Religious Knowledge to Global Problems.

CRS does not seek to prove or disprove religion. Its central purpose is to extract, model, and apply the knowledge embedded in religious texts to address pressing global challenges – whether in ethics, governance, conflict resolution, environmental stewardship, or social cohesion.

3. Operate Beyond the Constraints of Existing Social Science Frameworks.

While CRS may draw on insights from social sciences, it is not bound by their theoretical limits. The field is free to create its own conceptual models, theoretical assumptions, epistemic structures, and research methodologies, tailored to the unique nature of computationally modeling religious systems.

4. Treat Theological Foundations as Legitimate Modeling Variables.

CRS embraces the use of theological first principles – whether axioms, constants, or derived doctrines – as valid elements in computational models. There is no obligation to exclude such variables merely because they originate in revelation rather than empirical observation.

5. Empower Religion Without Engaging in Apologetics.

The goal of CRS is to equip religious traditions with modern analytical tools that amplify their relevance and applicability, not to engage in defensive apologetics. Empowerment means enabling religion to participate in contemporary intellectual and problem-solving arenas with methodological parity to other disciplines.

6. Develop Models That Can Challenge and Expand Existing Paradigms.

CRS aims to produce models that do more than supplement existing social science frameworks – they should be capable of challenging those models, offering alternative structures, and introducing new categories of analysis grounded in religious epistemology.

7. Foster Constructive Comparative Dialogue Between Traditions.

When conducting comparative studies between religions, the goal is not to declare one tradition correct and another mistaken. Rather, CRS provides a structured basis for constructive dialogue, mutual understanding, and cross-pollination of ethical and conceptual insights.

8. Allow for Multiple Modeling Interpretations of the Same Text.

Recognize that the same scriptural verse may yield multiple valid models depending on the interpretive lens, purpose, or contextualization. CRS remains open to diverse modeling approaches, provided they remain faithful to the semantic and doctrinal integrity of the text.

9. Establish and Use a Common Variable Dictionary.

For CRS to function as a collaborative discipline, it requires a shared lexicon of variables and semantic definitions. This book introduces the foundational variable dictionary for the field. Standardized semantic interpretations will ensure that researchers can reference each other's work with clarity and consistency.

10. Preserve the Original Semantic Integrity of Scriptural Language.

All CRS analysis is rooted in the actual text of scripture in its original language – whether Arabic for the Qur'an or Hebrew and Greek for the Bible. Translational work must take care to preserve the semantic precision and theological meaning of the original, ensuring that computational models are built on an accurate linguistic foundation.

META-FRAMEWORK: THE INTELLECTUAL CONSTITUTION OF COMPUTATIONAL RELIGIOUS STUDIES (CRS)

Computational Religious Studies is not simply another subfield of religious studies, theology, or digital humanities. It is an independent discipline with its own philosophical stance, epistemic commitments, and methodological freedoms. The ten foundational principles outlined above are not procedural guidelines – they are the constitution of the field, defining its identity and protecting its integrity.

CRS begins with the conviction that religion is not a peripheral or outdated category of human knowledge, but a central repository of intellectual, ethical, and civilizational wisdom. It affirms that sacred texts are not merely historical artifacts or cultural expressions, but living systems of knowledge capable of being modeled, analyzed, and applied in ways that address humanity's most urgent challenges. In this respect, CRS stands apart from secular academic disciplines that may reduce religion to a purely sociological, psychological, or anthropological phenomenon.

By grounding itself in the original languages and semantic structures of scripture, CRS ensures that its computational models remain faithful to the source material's integrity. By embracing theological first principles as valid inputs, it asserts methodological independence from empirical-only frameworks without sacrificing analytical rigor. By promoting a common variable dictionary, it enables the discipline to function as a cumulative, collaborative science – where models can be built upon, refined, and tested across contexts and generations.

Importantly, CRS is not defensive in posture. It does not exist to shield religion from critique, nor to place it in competition with the sciences. Rather, it operates with a posture of *empowerment*: providing religion with the computational tools, modeling methodologies, and epistemic clarity to take an active role in shaping the intellectual and practical solutions of the modern world.

As such, CRS is both a reclamation and a reinvention. It reclaims the pre-modern understanding of theology as a *scientia* – a disciplined body of knowledge deduced from first principles – while reinventing it for the computational age. This reinvention does not dilute the faith commitments of a tradition; instead, it makes those commitments structurally intelligible and operationally applicable within modern interdisciplinary discourse.

In short, CRS stands as a field with its own epistemological scaffolding, its own methodological rules, and its own vision: to bring the enduring wisdom of religion into full engagement with the tools, challenges, and opportunities of the 21st century, without compromise to its textual integrity or theological depth.

CRS METHODOLOGICAL FLOW

CRS Methodological Flow is a step-by-step framework from scripture to computational model, preserving theological integrity while enabling rigorous analysis.

Step 1 – Define the Scope and Purpose

—— Clarify the objective: Identify the global, societal, or theoretical problem your study will address (e.g., conflict resolution, ethical governance, environmental stewardship).

—— Determine the religious tradition(s) involved and the specific domain of theological knowledge to be modeled (e.g., justice in Islamic theology, covenantal ethics in the Bible).

—— State intent in accordance with CRS principles: the work must empower religion, not seek to disprove it.

Step 2 – Source Identification and Textual Grounding

—— Begin with the original-language scripture (Arabic for the Qur'an, Hebrew/Greek for the Bible) to ensure semantic fidelity.

—— Identify primary sources (scripture) and secondary sources (classical commentaries, theological treatises, recognized exegetical authorities).

—— Document all sources in a way that preserves their interpretive lineage.

Step 3 – Semantic and Theological Analysis

—— Conduct a close reading of the selected text, focusing on the semantic range of key terms in the original language.

—— Identify theological first principles (axioms) relevant to the topic.

—— Extract derived propositions (doctrinal conclusions) from authoritative interpretations.

—— Preserve multiple valid interpretations when they exist, noting contextual or school-specific distinctions.

Step 4 – Variable and Constant Identification

—— Map theological principles into variables (concepts that can vary by context, e.g., "justice," "mercy," "human agency") and constants (unchanging truths, e.g., "God's omniscience").

—— Assign each a formal definition from the CRS Variable Dictionary for consistency.

—— Where no existing definition exists, propose one for peer review and eventual integration into the shared dictionary.

Step 5 – Model Design

—— Choose the modeling approach: logical model, systems theory map, network graph, agent-based simulation, or hybrid.

—— Maintain internal coherence: the model must preserve the theological reasoning pathways of the tradition, even when translated into formal logic or computational rules.

—— Make epistemic assumptions explicit: declare where the model assumes revelation-based truths versus where it engages external knowledge sources.

In CRS, model design begins with selecting the structural paradigm – the conceptual form that best reflects the theological reasoning you are modeling. This could be:

—— Logical Model – representing doctrinal reasoning as formal propositions and inference rules.

—— Systems Theory Map – depicting theological concepts as interrelated components in a dynamic system.

—— Network Graph – mapping relationships among theological variables or scriptural concepts as nodes and edges.

—— Agent-Based Model – simulating interactions among agents (human or conceptual) under theological constraints.

—— Hybrid – combining elements of multiple paradigms.

These structural paradigms define the architecture of your model. However, CRS does not stop at the conceptual stage – mathematical formalization is essential to move from abstract structure to an operational, testable model.

This is where foundational mathematical approaches – common in political science, social research, and applied mathematics – become indispensable:

—— Algebra for expressing direct relationships between theological variables.

—— Calculus for modeling continuous change, growth, or decay (e.g., rates of moral adoption or spiritual decline).

—— Probability and Statistics for representing uncertainty, stochastic processes, and probabilistic reasoning in doctrinal contexts.

—— Linear Algebra for encoding and analyzing networks or complex interdependencies.

— Differential Equations for simulating dynamic theological systems that evolve over time.
— Formal Logic Systems for preserving deductive and inferential integrity.

In other words, the structural paradigm is the skeleton; the mathematics is the musculature and connective tissue. For example:

— A network graph of Qur'anic concepts may be implemented as an adjacency matrix (linear algebra) and analyzed using centrality measures.
— An agent-based model of moral decision-making may integrate probability theory to simulate varied responses under theological constraints.
— A systems theory map of eschatological events could be expressed as a set of coupled differential equations to model sequencing and timing under different interpretations.

CRS researchers must therefore design their models at two levels:

— Architectural – choosing the paradigm that mirrors the theological reasoning pattern.

— Mathematical – selecting and applying the right mathematical tools to formalize, execute, and test the model.

By operating at both levels, CRS ensures that models are faithful to the internal logic of the tradition while also being computationally precise and analytically powerful.

Step 6 – Simulation and Testing

— Implement the model in a computational environment (e.g., rule-based AI, agent simulation, mathematical system).

— Run scenario analyses to explore how theological reasoning responds to different contextual inputs.

— Preserve the integrity of the tradition's logic – no scenario should distort the original doctrinal reasoning.

Step 7 – Comparative and Dialogical Application

— When comparing with another tradition, use parallel modeling structures to ensure fairness and preserve each tradition's epistemic frame.

— The purpose is constructive dialogue, not adjudicating correctness.

— Highlight areas of conceptual overlap, divergence, and potential complementarity.

Step 8 – Interpretive Feedback Loop

—— Compare computational outputs with classical theological conclusions.
—— Consult domain scholars to validate whether the model's reasoning aligns with the tradition's integrity.
—— Revise variables, rules, or relationships to correct any distortions introduced by computational translation.

Step 9 – Documentation and Knowledge Integration

—— Fully document the model's structure, variables, and assumptions.
—— Publish models in a way that is reproducible by other CRS researchers.
—— Contribute new variables or revised definitions to the CRS Variable Dictionary to expand the field's shared resources.

Step 10 – Ethical and Faithful Application

—— Apply the model's insights to real-world contexts, policy discussions, or interdisciplinary research in a way that empowers religion as an equal epistemic partner.
—— Ensure that the application does not distort, trivialize, or decontextualize sacred teachings.
—— Reflect on potential unintended consequences before deployment in public, policy, or technological contexts.

OUR JOURNEY

In this book, the reader will embark on a transformative journey through the emerging field of Computational Religious Studies (CRS), an interdisciplinary approach that bridges sacred wisdom with modern scientific and technological tools. The journey begins with Part I: Foundations of a New Discipline, where the core concepts of the Computational Quran are introduced. In Chapter 1: Computational Religious Studies (CRS): The Birth of a New Field introduces the theoretical foundation of CRS, distinguishing it from traditional digital humanities and showcasing its potential to unify theology, philosophy, and science. Chapter 2: What Is the Computational Quran?, the book explains how sacred texts, particularly the Quran, can be treated as a structured system of ethical, cognitive, and behavioral knowledge that is not just readable by humans but also intelligible to artificial intelligence systems. This lays the groundwork for understanding how scripture can be modeled using systems theory, mathematical logic, and computational tools. Chapter 3: Scripture as System takes the reader deeper into the nature of religious scripture as a dynamic system that can be modeled, with an emphasis on how sacred texts contain inputs, outputs, and relational structures that transcend linear interpretation.

In Part II: Methodology and Mathematical Framework, the reader will explore the technical framework for modeling the Quran computationally. Chapter 4: The Five-Step Modeling Framework introduces a systematic process for transforming Quranic verses into computationally analyzable forms. Chapter 5: Mathematical Techniques in Scripture Modeling details the mathematical and logical tools used to model the Quran, including calculus, graph theory, and belief propagation. Chapter 6: The Universal Quranic Variables introduces the Universal Quranic Variable Ontology (UQVO), a standardized framework that enables both human and machine interpretations of the Quran.

Part III: Structural Modeling of the Quran applies these methods to real Quranic content. In Chapter 7: Modeling Surahs and Verses, the book explores how the opening Surah (chapter 1) and 40 verses of the second chapter of the Quran establishes the foundational relationship between humanity and the Divine. Chapter 8: Applied CRS takes a deeper dive into how to use mathematical models with AI, algorithmic reasoning, AI agents, and simulations. Chapter 9: examines paths to learn, research, get certified, and participate in the CRS revolution. Chapter 10 continues with the themes developed in Chapter 9 to show how scholars, researchers, and practitioners can get involved.

The final section, Part IV: Applications, Implications, and Future Directions, explores how CRS can be integrated into real-world applications. Chapter 11: From Scripture to Machine: Integrating the Computational Quran with AI looks at how Quranic models can guide AI systems, especially in ethical reasoning and spiritual intelligence. Chapter 12 discusses advances in knowledge representation and the use of Computational Quran in the quantum era. Chapter 13: CRS vs Interpretation: Solving the Hermeneutics Problem presents CRS as a solution to the often subjective and fragmented nature of scriptural interpretation. Chapter 14: CRS and Comparative Religion extends this modeling logic to comparative studies, setting the stage for the *Computational Bible*. The journey concludes in Chapter 15: Toward a New Science of Revelation, where the book proposes a new model of religious epistemology – one that integrates divine knowledge with modern scientific inquiry and artificial intelligence.

By the end of this book, readers will not only have a deeper understanding of how the Quran can be structured, analyzed, and applied in the modern world but will also be part of a larger movement that redefines the intersection of faith, technology, and knowledge. This journey will provide a new lens through which to explore sacred texts, offering innovative ways to integrate religious wisdom into the technological advancements of the future.

REFERENCES

al-Shahrastānī, M. (1923). *Kitāb al-Milal wa al-Nihal* [The book of sects and creeds]. Cureton, W., Ed.). Otto Harrassowitz.

Aquinas, T. (1920). *Summa theologica* (Fathers of the English Dominican Province, Trans.). London, England: Burns, Oates & Washbourne. (Original work published ca. 1265–1274)

Asad, T. (1993). *Genealogies of religion: Discipline and reasons of power in Christianity and Islam*. Johns Hopkins University Press. DOI: 10.1353/book.16014

Byrne, P. (n.d.). *Prolegomena to the philosophy of religion*.

Campbell, H. (2007). Who's got the power? Religious authority and the Internet. *Journal of Computer-Mediated Communication, 12*(3), 1043–1062. DOI: 10.1111/j.1083-6101.2007.00362.x

Chryssides, G. D., & Geaves, R. (2007). *The study of religion: An introduction to key ideas and methods*. Continuum.

Downey, A. B. (2014). *Religious affiliation, education and Internet use*. https://arxiv.org/abs/1403.5534

Durkheim, É. (1995). *The elementary forms of religious life* (Fields, K. E., Trans.). Free Press. (Original work published 1912)

Eliade, M. (1959). *The sacred and the profane: The nature of religion* (Trask, W. R., Trans.). Harcourt.

Evans-Pritchard, E. E. (1937). *Witchcraft, oracles and magic among the Azande*. Clarendon Press.

Ford, D. F. (2005). *The modern theologians: An introduction to Christian theology since 1918* (3rd ed.). Blackwell.

Frazer, J. G. (1890). *The golden bough: A study in magic and religion*. Macmillan.

Freud, S. (1913). *Totem and taboo*.

Freud, S. (1961). *The future of an illusion* (Strachey, J., Trans.). W. W. Norton. (Original work published 1927)

Kerr, A. (2002). The Temple of Jesus' Body.

Kritzeck, J. (1964). *Peter the Venerable and Islam*. Princeton University Press. DOI: 10.1515/9781400875771

Krügel, S., Ostermaier, A., & Uhl, M. (2023). ChatGPT's inconsistent moral advice influences users' judgment. *Scientific Reports, 13*(1), 4569. Advance online publication. DOI: 10.1038/s41598-023-31341-0 PMID: 37024502

Lane, J. E., & Shults, F. L. (2021). The computational science of religion. *Journal for the Cognitive Science of Religion, 6*(1-2). Advance online publication. DOI: 10.1558/jcsr.38669

McClure, P. K. (2020). The buffered, technological self: Finding associations between Internet use and religiosity. *Social Compass, 67*(3), 461–478. DOI: 10.1177/0037768620922128

McGrath, A. E. (2017a). *Christian theology: An introduction* (6th ed.). Wiley-Blackwell.

McGrath, A. E. (2017b). *Theology: The basics* (4th ed.). Wiley-Blackwell. DOI: 10.1002/9781119158110

McGrath, A. E. (2001–2003). *A scientific theology* (Vols. 1–3). Grand Rapids, MI: Eerdmans.

Morreall, J. H., & Sonn, T. (2011). *The religion toolkit: A complete guide to religious studies*. Wiley-Blackwell.

Müller, F. M. (1873). *Introduction to the science of religion*. Longmans, Green.

Müller, F. M. (1878). *Lectures on the origin and growth of religion*. Longmans, Green.

Orsi, R. A. (2003). Is the study of lived religion irrelevant to the world we live in? *Journal for the Scientific Study of Religion, 42*(2), 169–174. DOI: 10.1111/1468-5906.t01-1-00170

Otto, R. (1923). *The idea of the holy: An inquiry into the non-rational factor in the idea of the divine and its relation to the rational* (Harvey, J. W., Trans.). Oxford University Press.

Özcan, Ş. (2019). Ancient Origins of the History of Religions: Herodotus Example. *İlahiyat Tetkikleri Dergisi*, (51), 395-417.

Pyysiäinen, I. (2013). Cognitive science of religion: State-of-the-art. *Journal for the Cognitive Science of Religion, 1*(1), 5–28. DOI: 10.1558/jcsr.v1i1.5

Sharpe, E. J. (1986). *Comparative religion: A history* (2nd ed.). Duckworth.

Smart, N. (1998). *The world's religions* (2nd ed.). Cambridge University Press.

Tylor, E. B. (1871). *Primitive culture* (Vol. 1–2). John Murray.

Chapter 2
The Upcoming AI Is Cognitive

ABSTRACT

This chapter argues that the next generation of artificial intelligence will move beyond generative, pattern-recognition systems toward cognitive machines built on world models - internal representations that track states, causes, consequences, and possible futures rather than merely predicting language. Drawing on research in world models, neurosymbolic AI, knowledge graphs, and joint embedding predictive architectures, the chapter distinguishes fluent linguistic output from grounded understanding and explains why this distinction is decisive for sacred texts. The authors contend that if scripture is supplied to AI only as raw text, future systems risk mistaking surface association for faithfulness, flattening moral hierarchy and losing the causal structure that revelation encodes. The Computational Quran is presented as a corrective: a means of representing Quranic guidance as explicit agents, states, transitions, constraints, and consequences suitable for cognitive architectures.

INTRODUCTION

Right now, many of us are already using Artificial Intelligence to do amazing things. GenAI is helping us analyze papers, write content, edit ideas, and move faster in our work. Generative AI is powerful, and that is why it feels so exciting. But this is only the beginning. The next generation of machines will go even further, becoming more capable, more intelligent, and more truly cognitive in how they understand and respond to the world.

Today, Large Language Models (LLMs) are being developed to capture emotional dimensions (Ratican 2023, Lee 2023), provide moral reasoning (Zhao 2022), show emotional intelligence (Wang 2023), and display personalities (Gu 2023).

DOI: 10.4018/979-8-3373-6940-2.ch002

Fundamentally, LLMs remain pattern-recognition systems: they generate outputs by learning statistical relationships in data rather than by understanding the world as humans do. This is powerful, but it should not be confused with full intelligence. The next frontier may move closer to natural intelligence through systems that build world models, reason over causes and consequences, and act with memory, goals, and context.

FROM PATTERN RECOGNITION TO COGNITIVE MACHINES: WHY WORLD MODELS MATTER FOR THE COMPUTATIONAL SCRIPTURES

Today's artificial intelligence is astonishingly capable, but its capabilities must be understood carefully. Much of modern AI, especially large language models, is extraordinarily good at pattern recognition. It can predict words, summarize long documents, write code, translate languages, and find hidden relationships in large datasets. These are powerful achievements. Yet they do not necessarily amount to full cognition. A system may sound intelligent because it has learned from vast amounts of text, but fluency is not the same as understanding, judgment, or wisdom.

The basic limitation is that large language models are primarily trained on linguistic form. They learn patterns in sequences of words and produce likely continuations. Bender and Koller famously argue that a system trained only on form has no direct access to meaning in the grounded human sense (Bender & Koller, 2020). It may manipulate language impressively, but it does not necessarily understand the world that language describes. This is why AI systems can produce confident but false answers. TruthfulQA, a benchmark developed by Lin, Hilton, and Evans, showed that language models can imitate human falsehoods and generate misleading answers, especially when those falsehoods are common in training data (Lin et al 2022).

This does not mean today's AI is useless. It means that its intelligence is incomplete. A language model can answer, "What is justice?" by drawing on millions of texts. But a cognitive machine would need to go further. It would need to represent justice as a principle, observe when justice is violated, understand the agents and institutions involved, predict consequences, compare moral alternatives, and choose actions under constraints. That requires more than language. It requires a model of reality.

This is where world models become central. A world model is an internal representation of how a domain works. It is like a mental simulator. Instead of merely asking, "What is the most likely next word?" it asks, "What state am I observing? What caused it? What may happen next? What action should be taken?" Ha and Schmidhuber's influential paper, "World Models," showed that an AI agent could learn a compressed spatial and temporal representation of an environment and use it to guide action (Ha and Schmidhuber 2018). Their work demonstrated that an agent can train not only by reacting to the world, but also by using an internal model of the world to imagine possible futures.

Yann LeCun has made a similar argument about the future of AI. In "A Path Towards Autonomous Machine Intelligence," he argues that advanced intelligence will require predictive world models, memory, planning, self-supervised learning, and hierarchical action (LeCun 2022). The reason is straightforward: real intelligence is not merely the ability to describe reality. It is the ability to anticipate reality, act within it, and adapt over time.

This is why future cognitive machines may appear far smarter than today's AI systems. They may remember situations across time, understand causal chains, simulate consequences before acting, and optimize decisions for long-term goals. A current AI chatbot can give a fluent answer about a crisis. A cognitive machine would model the crisis as a changing system: agents, incentives, risks, constraints, values, possible interventions, and likely consequences.

Neurosymbolic AI is another important part of this future. Neural systems are excellent at detecting patterns in messy data: language, images, audio, signals, behavior, and documents. Symbolic systems are better at rules, logic, constraints, explicit reasoning, and traceable structure. Neurosymbolic AI tries to combine both. Besold and colleagues describe neural-symbolic learning and reasoning as an effort to integrate machine learning with computational logic and automated reasoning (Besold et al., 2017). A more recent survey similarly describes neural-symbolic systems as combining the perception strength of neural networks with the reasoning strength of symbolic systems (Ji et al., 2022).

This matters because LLMs alone are not enough for full cognition. They struggle with stable understanding, long-term planning, consistency, and explicit reasoning over time. They can generate plausible answers without maintaining a coherent model of the underlying situation. The future of AI will therefore likely combine several elements: language models, world models, memory, planning systems, symbolic reasoning, agents, and feedback loops (Garcez et al. 2019). In the language of cognition, future systems will need to sense, reason, act, and learn.

This is where the Computational Quran becomes important. The Quran is not merely a collection of statements. It presents a world: a moral universe of agents, intentions, actions, consequences, guidance, accountability, mercy, justice, corruption, repentance, arrogance, and social order. If this world is given to AI only as raw text, the system may learn surface associations. It may quote verses, summarize commentaries, or imitate theological language. But if Quranic guidance is represented computationally – as variables, relationships, states, transitions, constraints, and ethical principles – it becomes more suitable for future cognitive machines.

For example, a text model can say that oppression is condemned. A computational model can represent oppression structurally: an agent misuses power; a vulnerable party is harmed; moral limits are violated; warning is ignored; accountability becomes probable; repentance or reform may alter the trajectory. This allows a world model to learn not just the sentence, but the moral-causal structure behind the sentence.

That is the deeper promise of the Computational Quran. It can help express Quranic meaning in forms that future AI systems can reason over more transparently. It can convert interpretive claims into explicit structures that can be examined, compared, and challenged. It can help move religious AI from quotation to cognition.

The future of AI will not be defined only by machines that speak well. It will be defined by machines that model reality well. World models matter because they move AI from fluent response to structured understanding from deeper latent representations (Assram et al. 2023). And if future machines require structured ethical knowledge, then the Computational Quran may become one of the most important bridges between sacred guidance and cognitive architecture.

WORLD MODELS: THE NEXT STEP BEYOND LANGUAGE MODELS

A world model is an AI system that learns an internal representation of how some part of the world works. Instead of merely producing words, it tries to model states, causes, consequences, and possible futures. In simple terms, a world model is an AI's "mental simulator." It observes the environment, learns patterns of change, predicts what may happen next, and helps an agent choose actions before actually taking them.

The idea is not new. In a widely cited paper titled "World Models," David Ha and Jürgen Schmidhuber showed that an AI agent could learn a compressed representation of an environment and then use that internal model to act effectively (Ha and Schmidhuber 2018). Their system could even train inside its own simulated "dream" of the environment before transferring behavior back to the real task. The key insight was profound: intelligence is not only about reacting to the world; it is about building an internal model that allows prediction and planning.

This is different from a large language model, or LLM. An LLM, such as GPT-style systems, is primarily trained to predict the next token in a sequence of text. It learns statistical relationships among words, phrases, documents, and concepts. This makes it extremely powerful for language tasks: writing, summarizing, coding, answering questions, and reasoning over text. But the world is not made of text alone. The world is made of objects, forces, people, incentives, signals, movement, time, risk, and consequences. LLMs can describe these things, but they do not necessarily possess a grounded model of how they unfold in reality.

A world model, by contrast, aims to learn the structure of reality itself, or at least a useful approximation of it. It may learn from images, video, sensor streams, robot actions, battlefield signals, supply chain flows, financial movements, or social behavior. Its goal is not simply to say the next likely sentence. Its goal is to answer questions such as: What state am I in? What caused this? What will happen next? What happens if I act? What should I do to reach a goal?

Yann LeCun has argued that this kind of architecture is essential for future autonomous intelligence. In "A Path Towards Autonomous Machine Intelligence," he proposes systems built around predictive world models, intrinsic motivation, hierarchical planning, and self-supervised learning. His argument is that truly capable AI agents need to learn internal models of the world so they can reason, plan, and adapt over multiple time horizons (LeCun 2022).

This is also why JEPA, or Joint Embedding Predictive Architecture, is important. Rather than predicting every pixel or every word, JEPA-style systems predict representations in a latent space. The Image-based JEPA paper showed that models can learn useful semantic representations by predicting missing parts of an image in representation space, not by reconstructing superficial details (Assram et al. 2021). This matters because intelligent agents do not need to remember every detail of the world. They need to understand what is relevant: objects, relationships, constraints, risks, affordances, and likely next states.

World models are likely to be the future because the next major frontier of AI is not conversation; it is action. We need AI that can help robots move safely, vehicles drive reliably, commanders understand dynamic battlefields, factories anticipate failures, hospitals manage patient trajectories, and governments simulate policy consequences. These tasks require models of changing systems, not merely models of language.

LLMs will remain important, but they are best understood as one component of a larger cognitive architecture. They are powerful interfaces for language, explanation, and symbolic reasoning. World models are more suitable for prediction, planning, control, and adaptation. The future will likely combine both: LLMs to communicate and reason in human language, and world models to understand dynamic reality.

The simplest distinction is this: an LLM predicts what words should come next; a world model predicts what state of the world may come next. That difference is enormous. Human intelligence is not merely linguistic. We imagine futures, test actions mentally, learn from consequences, and adjust behavior. World models attempt to give AI systems that same basic capacity: not just to talk about the world, but to anticipate and act within it.

FROM SCRIPTURE TO WORLD MODELS: WHY SACRED TEXTS NEED COMPUTATIONAL REPRESENTATION

Piotrowski shows that historical texts require specialized computational treatment because their linguistic instability, spelling variation, and textual transmission issues make ordinary NLP methods insufficient (Piotrowski 2012). However, Daquino, Pasqual, and Tomasi argue that hermeneutical claims can be represented through Semantic Web technologies, making interpretation more portable, comparable, assessable, and reusable (Daquino et al. 2020).

The central challenge of interpreting scripture has never been the absence of words. The words are present, preserved, recited, translated, debated, and studied across centuries. The challenge is that sacred texts contain layers of meaning: moral principles, legal judgments, social models, divine commands, spiritual states, human motivations, historical examples, and consequences for action. These meanings are not always visible from the surface of the text alone. They require interpretation.

For most of history, interpretation has been conducted through human scholarship: grammar, commentary, law, theology, philosophy, history, and spiritual reflection. This work remains essential. But the rise of artificial intelligence introduces a new possibility. Scripture can now be represented not only as text, but also as structure: a system of variables, relationships, states, transitions, causes, consequences, and governing principles. This is the foundation of what may be called Computational Scripture.

Computational Scripture does not reduce revelation to mathematics. It does not claim that sacred meaning can be exhausted by equations. Rather, it seeks to make explicit the internal structure of interpretive claims. When a verse speaks about justice, mercy, oppression, repentance, arrogance, covenant, authority, or accountability, the computational question becomes: What are the entities? What are the relationships? What state is being described? What action changes that state? What principle governs the transition? What consequence follows? What conditions or exceptions matter?

This movement from text to structure is crucial because the next generation of AI will not be built only around language models. It will increasingly be built around world models.

A large language model, or LLM, is primarily trained to predict language. It learns patterns in enormous bodies of text and becomes extraordinarily capable at producing fluent responses, summaries, arguments, translations, and explanations. But language prediction is not the same as modeling the world. Scholars such as Emily Bender and Alexander Koller have argued that systems trained only on linguistic form do not necessarily acquire grounded meaning in the human sense (Bender &

Koller, 2020). They may manipulate symbols with great fluency while lacking a structured understanding of the realities those symbols refer to.

A world model has a different purpose. It attempts to learn how a domain works. It represents states, relationships, changes, possible futures, and consequences. David Ha and Jürgen Schmidhuber's influential paper, "World Models," showed how an AI agent could learn an internal compressed representation of an environment and use it to guide action. Yann LeCun has similarly argued that future autonomous intelligence will require predictive world models, hierarchical planning, and internal representations of how reality changes over time.

The simplest distinction is this: an LLM predicts the next word; a world model predicts the next state.

This distinction matters deeply for scripture. Scripture is not merely a sequence of statements. It is filled with state transitions. A person moves from heedlessness to awareness, arrogance to humility, sin to repentance, injustice to accountability, despair to hope, gratitude to guidance. Communities rise, decay, reform, or collapse. Leaders become just or tyrannical. Wealth becomes charity or corruption. Knowledge becomes wisdom or pride. Power becomes stewardship or oppression.

These are not simply literary themes. They are moral-causal patterns. Scripture repeatedly shows that certain inner states produce certain actions, that certain actions reshape individuals and communities, and that certain consequences follow from moral choices. In this sense, scripture already contains a kind of world model: a model of human beings, society, authority, temptation, justice, mercy, responsibility, and destiny.

The problem is that ordinary text-based AI may not learn this model reliably. It may learn that certain words are associated with certain interpretations. It may summarize classical commentaries. It may reproduce doctrinal positions. It may compare schools of thought. But unless the underlying structure is represented explicitly, the model remains vulnerable to surface-level associations, rhetorical confidence, and inconsistent reasoning.

Computational representation changes this. It turns implicit meaning into explicit structure. A scriptural passage can be modeled as a set of agents, states, actions, constraints, and outcomes. For example:

$$S_{t+1} = f(S_t, A_t, C_t, G_t, R_t)$$

Here, represents the current moral, social, or cognitive state; represents human action; represents context; represents divine guidance or governing principle; and represents response, consequence, or accountability. The formula does not replace the verse. It clarifies the structure of the interpretation.

This is where knowledge graphs and neuro-symbolic AI become important. Knowledge graphs represent entities and relationships explicitly. They allow a system to understand that a prophet warns a community, that a ruler oppresses the weak, that wealth creates obligations, that testimony requires evidence, that justice constrains authority, and that mercy interacts with accountability. Survey work by Ji and colleagues has shown how knowledge graphs support representation learning, reasoning, and knowledge-aware AI systems (Ji et al., 2022). Neural-symbolic AI research, reviewed by scholars such as Besold and others, has similarly argued that future intelligent systems benefit from combining statistical learning with explicit symbolic structures (Besold et al., 2017).

For scripture, this combination is essential. Interpretation is not just pattern recognition. It involves rules, exceptions, analogies, objectives, hierarchy of principles, historical context, moral priorities, and contested authority. A purely text-based model may generate plausible commentary. A computationally structured model can expose the logic behind the commentary.

This is especially important because sacred texts are often used in radically different ways. The same text may be invoked for liberation or domination, mercy or cruelty, justice or authoritarianism. Without an explicit methodological structure, interpretive claims often compete through rhetoric, authority, or inherited tradition. Computational Scripture offers another layer: interpretive claims can be represented structurally and compared. What assumptions are being made? Which principle is treated as dominant? Which context is being included or excluded? Which agent has authority? Which consequence is predicted? Which moral variable is being ignored?

This does not eliminate disagreement. But it makes disagreement more transparent.

The connection to modern world-model research is direct. Recent approaches such as Joint Embedding Predictive Architectures, or JEPA, show that AI systems can learn by predicting abstract representations rather than reconstructing surface-level data. The lesson is powerful: intelligence may depend less on memorizing surface details and more on learning the deeper structure of reality. Computational Scripture follows the same logic. Instead of asking AI only to predict the next word in a verse, commentary, or translation, we can ask it to model the next moral state, the governing principle, the likely consequence of an action, or the tension between competing values.

Consider the figure of Pharaoh in the Qur'an and the Bible. A text-only model may learn that Pharaoh is associated with tyranny because many texts say so. A computational world model can go further. It can represent Pharaoh as a recurring structure: concentrated power, epistemic arrogance, oppression of the vulnerable, rejection of warning, confrontation with prophetic truth, delayed accountability, and eventual collapse. Once represented structurally, this pattern can be generalized. It can be examined in rulers, institutions, corporations, empires, or even individual psychology.

That is the difference between memorizing a story and modeling a moral universe.

The argument must be stated with care. It is not yet empirically proven that a fully developed Computational Scripture system will outperform text-only models across all interpretive tasks. That remains to be tested. But the direction is strongly supported by existing AI research. LLMs are powerful but limited as text-prediction systems. World models are designed to learn states, transitions, and consequences. Knowledge graphs and neuro-symbolic systems improve explicit reasoning. Computational hermeneutics shows that interpretation itself can be represented and studied through computational methods.

The future of AI and scripture, therefore, is not simply building chatbots that answer religious questions. That is only the first stage. The deeper opportunity is to build structured moral-cognitive models that allow AI systems to reason more transparently about scriptural worlds: the world of the soul, the world of society, the world of power, the world of justice, and the world of divine accountability.

Computational Scripture is not merely digitized scripture. It is scripture represented as moral architecture. It converts sacred text into explicit structures of agency, responsibility, causality, and consequence. Such structures are far more suitable for world models than raw text alone because world models learn how states change, how actions matter, and how principles constrain possible futures.

An LLM can speak about scripture. A scripture-informed world model could begin to model the moral universe that scripture describes.

PATTERN-BASED AI IS DANGEROUS FOR SCRIPTURES, COGNITIVE IS WORSE IF NOT APPROACHED CORRECTLY

As Cognitive AI begins to emerge, people of faith must take responsibility now. They cannot wait until these systems are fully built and already shaping how millions think and learn. If future intelligent machines are going to form deep internal models of the world, then believers should help ensure those models reflect the Quran or

Bible with faithfulness, depth, and respect. We must help shape how these systems understand, reason about, and respond to Quranic (or Biblical) guidance. That work should begin now, before shallow or distorted models take hold. Below we present some of the reasons why:

1. Pattern-based AI can sound convincing without truly understanding scripture. The danger is not that AI will know nothing about the Quran. The danger is that it may appear to understand the Quran while actually relying on weak patterns from data.
2. Large language models learn from mixed and uneven sources. They pull from translations, commentaries, quotations, online discussions, polemics, summaries, and internet noise. All of this gets blended together, which may create fluent answers but not faithful understanding.
3. This becomes more serious as AI becomes more advanced. A future cognitive machine may build internal models of beliefs, values, behavior, and ethics. If its model of the Quran is based only on scattered patterns, then its internal understanding of Islam may become distorted, incomplete, or biased.
4. Pattern AI often loses important distinctions that scripture depends on. It may blur the difference between literal and contextual meaning, universal principles and case-specific instructions, moral ideals and legal rules, warnings and commands, or consensus and minority opinions. This makes the machine sound informed while actually mixing categories that should remain separate.
5. It is also vulnerable to bias in the digital environment. Online material about Islam is not evenly balanced or carefully weighted. Some views are overrepresented, others are underrepresented, and weak or polemical sources may shape the AI more than careful scholarship.
6. Pattern systems are better at spotting associations than preserving causal structure. The Quran does not only present topics; it often shows how belief, intention, justice, gratitude, patience, repentance, and other states lead to outcomes over time. A pattern model may notice these ideas appear together, but that is not the same as understanding how they are connected.
7. Another major risk is flattening moral hierarchy. Not everything in scripture has the same level of importance. Some teachings are foundational, some are conditional, some are narrative, and some are secondary. Pattern AI may treat all of it as equal "content," which changes the shape of the text.
8. This matters even more when AI begins to act, not just answer. Future systems may be used in education, advising, mediation, policy support, faith-community tools, and knowledge retrieval. If the underlying model is weak, then the system may give oversimplified guidance, confuse ethics with law, exaggerate fringe themes, or blur mainstream teaching with marginal views.

9. Pattern AI often optimizes for plausibility, not fidelity. It may produce answers that sound balanced, elegant, and helpful, while still being structurally unfaithful to the text. In scripture, sounding good is not enough; the response must also preserve meaning and interpretive discipline.
10. More capable AI can actually make the problem worse if it is not grounded. A stronger cognitive system may infer links, priorities, or generalizations that should not be inferred. The smarter it seems, the more dangerous its mistakes can become.
11. There is also a subtle substitution risk. AI may become very good at modeling what Muslims say, how communities behave, and what debates happen online. But that is not the same as modeling the Quran itself. Social behavior around the text can wrongly replace the structure of the text.
12. Sacred texts require constraint, not just intelligence. In scripture, the issue is not only making the model more powerful. It is making sure the model is disciplined, traceable, and structurally faithful.
13. Interpretability is essential for trust. A user should be able to ask what concept, structure, or assumption led to a conclusion. Pattern systems usually cannot explain this clearly because their knowledge is spread across hidden statistical relationships.

This is why Computational Quran (CQ) matters. CQ is not meant to replace faith or traditional tafsir. It is meant to help preserve Quranic structure in computational form so that future AI systems do not reduce sacred knowledge to shallow patterns.

The deepest risk is civilizational, not just technical. The problem is not one wrong answer or one hallucination. The real danger is that future intelligent systems may shape millions of interactions using a weak internal model of Quranic knowledge, causing users to mistake fluency for faithfulness.

THEORETICAL LINEAGE

The Computational Quran framework is original in its specific ambition: to represent Quranic revelation as a structured, computable, moral-causal system capable of supporting interpretation, comparison, simulation, and eventually world-model development. Yet this originality does not emerge in isolation. It stands at the intersection of several mature scholarly traditions: AI value alignment, digital humanities, computational sociology, and systems theory. Acknowledging these fields strengthens the framework because it shows that CQ is not an isolated speculative project, but a new synthesis built upon existing intellectual foundations.

1. AI Value Alignment

AI value alignment asks how intelligent systems can be designed so that their behavior remains consistent with human goals, values, ethical constraints, and social expectations. This field is directly relevant to CQ because a scripture-based computational model is, in part, a structured representation of values. If future AI systems require explicit moral and normative grounding, then the question becomes: which values, represented how, under whose authority, and with what interpretive safeguards?

Amodei et al.'s "Concrete Problems in AI Safety" is foundational because it frames AI failure not only as technical error but as harmful behavior emerging from poorly specified objectives, reward hacking, unsafe exploration, and distributional shift (Amodei et al., 2016). Russell's work on human-compatible AI similarly argues that the core AI problem is not raw capability, but ensuring that powerful systems act in accordance with human values and uncertainty about those values (Russell 2019). CQ contributes to this lineage by asking whether sacred moral systems can be represented in a form precise enough to inform AI reasoning without collapsing into simplistic rule lists.

2. Digital Humanities

Digital humanities provides the methodological bridge between classical textual scholarship and computational representation. It has long studied how historical, literary, and religious texts can be digitized, encoded, annotated, searched, visualized, and analyzed (Gold 2019). The edited volume *A Companion to Digital Humanities* is a major reference point because it frames digital humanities as more than digitization; it includes text encoding, scholarly infrastructure, computational methods, and new ways of producing and disseminating knowledge (Schreibman et al., 2004).

CQ extends this tradition. Digital humanities often asks: How can texts be preserved, encoded, searched, and analyzed computationally? CQ asks a further question: How can sacred texts be represented as structured systems of agents, principles, states, transformations, obligations, consequences, and interpretive constraints? In that sense, CQ is not a rejection of digital humanities. It is a next-stage development from textual computing toward epistemic and moral-causal modeling.

3. Computational Sociology

Computational sociology and computational social science (Cioffi-Revilla 2017) examine how large-scale data, networks, simulations, and computational methods can reveal patterns in individual and collective behavior. Lazer et al.'s landmark paper in *Science* describes computational social science as an emerging field enabled by the capacity to collect and analyze data at scales that reveal patterns of individual and group behavior (Lazer et al., 2009).

This lineage matters because scripture is not only concerned with individual belief. It repeatedly addresses communities, leadership, injustice, social fragmentation, wealth, power, conflict, moral contagion, and collective decline or reform. CQ therefore has a natural connection to computational sociology: it can model religiously described social dynamics as structured agent-state systems. For example, Quranic accounts of Pharaoh, corrupt elites, prophetic communities, hypocrisy, oppression, charity, and justice can be represented not merely as narratives, but as social models involving power, legitimacy, resistance, accountability, and transformation.

4. Systems Theory

Systems theory provides the deepest conceptual foundation for CQ. It treats reality as composed of interacting parts whose behavior cannot be fully understood by analyzing each part in isolation. Ludwig von Bertalanffy's general system theory sought common principles across biological, social, and technical systems, emphasizing interaction, organization, feedback, and wholeness (Bertalanffy 1968).

This is directly aligned with the CQ approach. The Quranic world is not a list of disconnected verses. It is a system of interrelated concepts: belief, action, intention, guidance, knowledge, arrogance, mercy, justice, repentance, oppression, community, law, and consequence. A systems perspective allows these concepts to be modeled as interacting variables rather than isolated themes (Meadows 2008). It also supports the CQ claim that interpretation should examine structure: how one concept modifies another, how moral states transition, how feedback occurs, and how individual and collective outcomes emerge over time.

From this perspective, the Computational Quran can be viewed as a synthesis, not an isolated invention. From AI alignment, it inherits the problem of representing values for intelligent systems. From digital humanities, it inherits the discipline of computational textual scholarship. From computational sociology, it inherits methods for modeling agents, groups, institutions, and collective behavior. From systems theory, it inherits the concept of structured wholes, feedback loops, and dynamic state transitions.

Its originality lies in bringing these lineages together around revelation. CQ does not merely digitize the Quran. It does not merely build a chatbot for religious questions. It does not merely use AI to summarize tafsir. It proposes a structured computational representation of Quranic meaning that can support interpretation, comparison, falsifiability, reproducibility, simulation, and eventually world-model construction.

The Computational Quran belongs to a broader theoretical lineage that includes AI value alignment, digital humanities, computational sociology, and systems theory. Its originality lies not in rejecting these traditions, but in integrating them into a new framework for representing revelation as a structured moral-causal system. This lineage legitimizes the project academically while clarifying its distinctive contribution: the movement from sacred text as language to sacred text as computable moral architecture.

REFERENCES

Amodei, D., Olah, C., Steinhardt, J., Christiano, P., Schulman, J., & Mané, D. (2016). *Concrete problems in AI safety*. arXiv. /arXiv.1606.06565DOI: 10.48550

Bender, E. M., & Koller, A. (2020). Climbing towards NLU: On meaning, form, and understanding in the age of data. In *Proceedings of the 58th Annual Meeting of the Association for Computational Linguistics* (pp. 5185–5198). Association for Computational Linguistics. DOI: 10.18653/v1/2020.acl-main.463

Besold, T. R., d'Avila Garcez, A., Bader, S., Bowman, H., Domingos, P., Hitzler, P., Kühnberger, K.-U., Lamb, L. C., Lowd, D., Lima, P. M. V., de Penning, L., Pinkas, G., Poon, H., & Zaverucha, G. (2017). *Neural-symbolic learning and reasoning: A survey and interpretation*. arXiv. /arXiv.1711.03902DOI: 10.48550

Cioffi-Revilla, C. (2017). *Introduction to computational social science: Principles and applications* (2nd ed.). Springer. DOI: 10.1007/978-3-319-50131-4

d'Avila Garcez, A., Gori, M., Lamb, L. C., Serafini, L., Spranger, M., & Tran, S. N. (2019). Neural-symbolic computing: An effective methodology for principled integration of machine learning and reasoning. *Journal of Applied Logic*, *6*(4), 611–631.

Daquino, M., Pasqual, V., & Tomasi, F. (2020). Knowledge representation of digital hermeneutics of archival and literary sources. *JLIS.it, 11*(3), 59–76. DOI: 10.4403/jlis.it-12642

Gold, M. K., & Klein, L. F. (Eds.). (2019). *Debates in the digital humanities 2019*. University of Minnesota Press. DOI: 10.5749/9781452963785

Gu, H., & Chandrasegaran, S. (2023). *On the effectiveness of creating conversational agent personalities through prompting*. [Manuscript].

Ha, D., & Schmidhuber, J. (2018). *World models*. arXiv. /arXiv.1803.10122DOI: 10.48550

Ji, S., Pan, S., Cambria, E., Marttinen, P., & Yu, P. S. (2022). A survey on knowledge graphs: Representation, acquisition, and applications. *IEEE Transactions on Neural Networks and Learning Systems*, *33*(2), 494–514. DOI: 10.1109/TNNLS.2021.3070843 PMID: 33900922

Lazer, D., Pentland, A., Adamic, L., Aral, S., Barabási, A.-L., Brewer, D., Christakis, N., Contractor, N., Fowler, J., Gutmann, M., Jebara, T., King, G., Macy, M., Roy, D., & Van Alstyne, M. (2009). Computational social science. *Science*, *323*(5915), 721–723. DOI: 10.1126/science.1167742 PMID: 19197046

LeCun, Y. (2022). *A path towards autonomous machine intelligence* (Version 0.9.2). OpenReview. https://openreview.net/forum?id=BZ5a1r-kVsf

Lee, Y. K., Lee, I., Shin, M., Bae, S., & Hahn, S. (2023). *Chain of empathy: Enhancing empathetic response of large language models based on psychotherapy models.* [Manuscript].

Lin, S., Hilton, J., & Evans, O. (2022). TruthfulQA: Measuring how models mimic human falsehoods. In *Proceedings of the 60th Annual Meeting of the Association for Computational Linguistics (*Volume 1*: Long Papers)* (pp. 3214–3252). Association for Computational Linguistics. DOI: 10.18653/v1/2022.acl-long.229

Meadows, D. H. (2008). *Thinking in systems: A primer.* Chelsea Green Publishing.

Piotrowski, M. (2012). *Natural language processing for historical texts.* Morgan & Claypool Publishers., DOI: 10.1007/978-3-031-02146-6

Ratican, J., & Hutson, J. (2023). The six emotional dimension (6DE) model: A multidimensional approach to analyzing human emotions and unlocking the potential of emotionally intelligent artificial intelligence via large language models. *1*(1), 44–52.

Russell, S. (2019). *Human compatible: Artificial intelligence and the problem of control.* Viking.

Schreibman, S., Siemens, R., & Unsworth, J. (Eds.). (2004). *A companion to digital humanities.* Blackwell.

von Bertalanffy, L. (1968). *General system theory: Foundations, development, applications.* George Braziller.

Wang, X., Li, X., Yin, Z., Wu, Y., & Liu, J. (2023). *Emotional intelligence of large language models.* DOI: 10.1177/18344909231213958

Zhou, J., Hu, M., Li, J., Zhang, X., Wu, X., King, I., & Meng, H. (2022). *Rethinking machine ethics: Can LLMs perform moral reasoning through the lens of moral theories?* [Manuscript]. Top of Form Bottom of Form

Chapter 3
Scripture as System

ABSTRACT

This chapter reframes scripture, especially the Qur'an, as a dynamic, open, and complex adaptive system rather than a static text. Drawing on systems theory, it shows how revelation continually interacts with human actions, historical conditions, and societal change through inputs, outputs, and feedback loops. The chapter explores emergent meaning, nonlinear cause–effect relationships, thresholds, and functional mappings that govern moral and social outcomes. By formalizing these dynamics mathematically, Computational Religious Studies reveals scripture as an operational architecture of guidance whose variables, equations, and feedback structures can be modeled, simulated, and ultimately integrated into AI and modern analytical frameworks.

INTRODUCTION

In the study of systems theory, a *system* is understood as an interconnected set of elements, organized in such a way that they produce a pattern of behavior over time. Systems may be open or closed, static or dynamic, simple or complex. They can be cybernetic – regulated through feedback – or adaptive, changing in response to internal and external stimuli (Bertalanffy 1968, Wiener 1948, Meadows 2008).

Open systems exchange matter, energy, or information with their environment (Bertanlanffy 1968). Closed systems, in contrast, are isolated, with little or no interaction beyond their boundaries. Dynamic systems change states over time; static systems remain in equilibrium. Complex systems are composed of many interacting components whose behavior cannot be fully understood by examining each part in isolation. A Complex Adaptive System (CAS) adds another layer: it evolves through feedback, self-organization, and adaptation to its environment (Stacey, 2011; Holland, 1995; Mitchell, 2009, Cilliers, 1998).

DOI: 10.4018/979-8-3373-6940-2.ch003

When this conceptual apparatus is applied to religious scripture – particularly the Qur'an – it becomes clear that sacred texts can be modeled as dynamic, vibrant systems, and specifically as complex adaptive systems (Gadamer, 2004, Iser 1978). The Qur'an is not a static artifact confined to the moment of its revelation. Rather, it is a living system that continuously interacts with its environment: the lives, cultures, and social systems of those who engage with it (Saeed, 2006; Rahman 2009).

AN OPEN SYSTEM

Even though the text of the Qur'an is static, the Qur'an is not closed upon itself. It exists in a constant exchange of "inputs" and "outputs" with the world of its believers and beyond.

Inputs include the act of recitation, study, and interpretation; the intellectual frameworks of scholars; and the historical and cultural contexts in which it is read (Gadamer, 2004; Saeed, 2006; Rahman, 2009). New knowledge – whether from archaeology, science, or social change – enters this interpretive process as additional input, shaping how the Qur'an is understood. The interpretation mechanism of Qur'an therefore becomes one of the inputs. Even challenges from outside the faith community serve as inputs, prompting renewed study and deeper exploration.

Outputs are the beliefs, laws, ethical norms, and social practices that flow from Qur'anic interpretation (Rahman, 2009; Saeed, 2006). These outputs, in turn, feed back into the environment – shaping communities, influencing governance, guiding moral decisions, and at times even altering the course of civilizations. In this way, the Qur'an functions as a system whose outputs directly reshape the very world from which its inputs are drawn.

A DYNAMIC SYSTEM

Unlike static texts, the Qur'an's engagement with humanity is dynamic (Gadamer, 2004; Saeed, 2006; Rahman, 2009). Its guidance is eternal, yet the application of that guidance is in perpetual motion, adapting to the shifting realities of time and place.

Interpretive Change: Over centuries, the emphasis on particular verses has shifted in response to social, political, and cultural change. A passage on justice may resonate differently in a society recovering from conflict than in one experiencing stability and prosperity. The core revelation remains constant, but its interpretive weight can vary with circumstances.

Emergent Meanings: As believers confront novel challenges – bioethics, digital governance, artificial intelligence – they return to the Qur'an for guidance, drawing new insights from verses that have been read for centuries. These emergent understandings are not distortions but adaptive manifestations of the Qur'an's capacity to address all ages, as promised in 16:89 (*"...and We have sent down to you the Book as clarification for all things..."*).

THE QUR'AN AS A COMPLEX ADAPTIVE SYSTEM

Based upon the above-mentioned properties, the Qur'an exhibits the hallmark properties of a CAS – emergence, self-organization, adaptation, and nonlinearity (Stacey, 2011; Holland, 1995; Mitchell, 2009; Cilliers, 1998; Kauffman, 1993) – making it uniquely resilient and perpetually relevant.

Emergence: The Qur'an's impact transcends the sum of its individual verses. Through centuries of interaction between text and community, it has generated rich traditions of scholarship, diverse schools of thought, and global networks of practice. The Qur'an in isolation is divine text; the Qur'an in interaction with humanity is a living civilization.

Self-Organization: While formal religious institutions exist, much of the Qur'an's lived application emerges organically from communities. Norms of worship, ethical customs, and local jurisprudence often develop without centralized planning, evolving through the collective actions and understandings of believers.

Adaptation: The Qur'an's guidance is interpreted anew in every age, allowing it to address both enduring human questions and unprecedented modern problems. This is not a change in the text but a change in its operational context. The divine law remains constant, but its engagement adapts, ensuring survival and growth of the faith community.

Nonlinearity: Minor interpretive shifts can have disproportionate effects. A new emphasis on a particular verse or principle can inspire reform movements, shift public discourse, or even lead to the emergence of entirely new social structures. The Qur'an thus mirrors nonlinear systems where small perturbations can lead to large and unpredictable outcomes.

When viewed through the lens of systems theory, the Qur'an is not a static record of past revelation. It is the central component of a dynamic, open, and adaptive system – one that interacts with human society, evolves through interpretation, and maintains vitality through constant feedback and adaptation. This perspective is not a modern imposition upon the text; it is, rather, an analytical recognition of how the Qur'an has always functioned in practice.

By framing scripture as a Complex Adaptive System, CRS provides the foundation for modeling its variables, functions, and feedback loops in a way that honors both its divine origin and its living role in human civilization.

Figure 1.

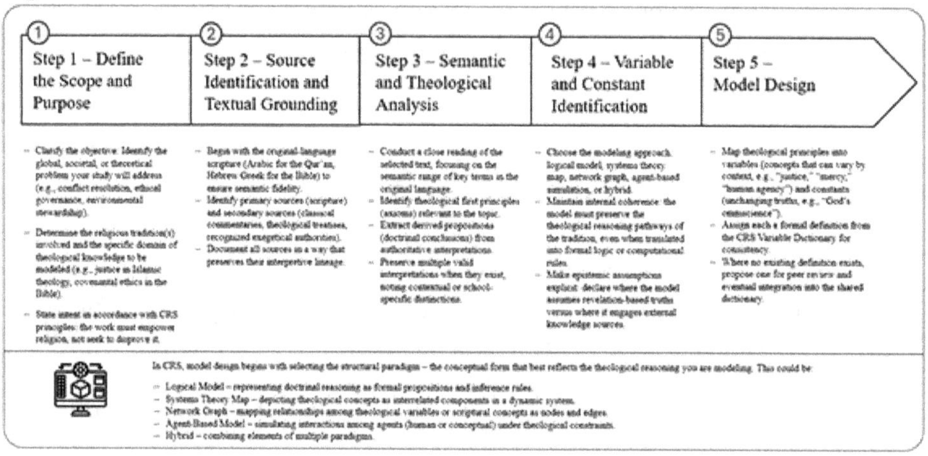

FROM TEXT TO SYSTEM

In Computational Religious Studies (CRS), scripture is not approached merely as a historical document, a collection of moral sayings, or a literary artifact. It is understood as an *operational system* – a divinely revealed architecture of principles, commands, and relationships designed to govern human conduct, shape social order, and guide civilizations toward their highest moral and spiritual states. This framing moves beyond passive reading to viewing scripture as a dynamic, structured mechanism with intentional design and systemic functionality.

The Qur'an itself affirms this perspective. It is introduced as *"a Book about which there is no doubt, a guidance for those conscious of God"* (2:2), affirming both certainty and purpose. Its verses are *"perfected in wisdom, then explained in detail from One who is All-Wise, All-Aware"* (11:1), underscoring the precision and coherence of its structure. Repeated reminders such as *"And We have certainly made the Qur'an easy for remembrance, so is there any who will remember?"* (54:17) highlight not only clarity of message but also the expectation of engagement, reflection, and application – features characteristic of an interactive system rather than a static record.

From a CRS perspective, the central thesis of this chapter is clear: If scripture is a system – intentionally structured, responsive to inputs, and generating transformative outputs – then it can be studied, analyzed, and modeled as such, without diminishing its divine authority.

Treating scripture as a system is not an act of reductionism; it is an act of precision. It allows us to capture, in formal terms, the logic, thresholds, and feedback loops that have always operated within revelation, thereby enabling deeper understanding and more effective application in a modern, computationally connected world.

In CRS, our objective is to identify general patterns among the abstract concepts presented in scripture. These concepts originate in divine revelation, transmitted from God to human cognition for the explicit purpose of guidance. Guidance therefore becomes the central organizing goal of the entire system.

However, once these revealed concepts enter human civilization, they cease to exist solely as abstract descriptions. They are applied in the fabric of daily life, shaping beliefs, behaviors, institutions, and laws. In this applied form, the concepts carry *specifications* and *expectations* for causal relationships – linking action to outcome, principle to result. These linked relationships form interconnected systems that operate within individual lives and across societies. Over time, they also generate an evidentiary record: observable effects and patterns that arise from adherence to, or deviation from, the divine prescriptions.

Such relationships can be expressed in the formal language of mathematics, where we work with defined terms, rules, and syntax to model how these systems operate. In CRS, our study is not *of mathematics itself* but *of scripture through mathematics*. Mathematics becomes a disciplined framework for conceptualization, a tool for building theories, specifying relationships, and testing hypotheses about the operation of divine guidance in human affairs.

By harnessing the formal power of mathematics, CRS transforms scripture from a solely textual source of guidance into a set of applied, practical frameworks for human life. This integration brings rigor to theoretical development and enables the systematic testing of a theory's implications. In doing so, it preserves the authority of revelation while extending its operational clarity, ensuring that divine guidance can be studied, understood, and applied with the same precision afforded to other complex systems in science and engineering.

THE NATURE OF SCRIPTURAL SYSTEMS

In systems theory, a *system* is generally defined as a set of interconnected components that interact to produce a pattern of behavior over time. Every system can be described in terms of its inputs, processes, outputs, and feedback.

When applied to the Qur'anic context, these elements take on specific meanings:

- *Inputs:* The system receives human actions, intentions, environmental conditions, and the prevailing state of society. These inputs include conscious acts of worship, moral choices, societal events, and the challenges of the era.
- *Processes:* Within the system, divine laws, moral logic, and the cause–effect relationships encoded in revelation govern how inputs are evaluated and transformed. This is where Qur'anic principles and commands operate, filtering human actions through a divinely set moral architecture.
- *Outputs:* The system generates consequences – both in the immediate and ultimate sense. These outputs include blessings and divine favor, punishments and warnings, societal stability or collapse, and transformations in the moral and spiritual state of individuals and communities.
- *Feedback:* The system is cybernetic, continually adjusting through feedback mechanisms – but always maintaining its divine guidance. Spiritual awareness, cycles of repentance, social reform, and collective moral awakening feed back into the system as new inputs, shaping the next iteration of the process.

  The Qur'an functions not as a linear system where inputs map to outputs in direct proportion, but as a nonlinear system. Outcomes are influenced by threshold effects – points at which a small change in behavior produces a disproportionate shift in consequence. This nonlinearity also reflects divine discretion: two identical outward actions may result in different divine responses due to unseen factors such as intention, sincerity, or broader societal impact.

In this light, the Qur'an emerges as a complex, adaptive system – one that continuously interacts with the human environment, processes inputs through a divine moral framework, and produces outputs that in turn reshape the very context from which the next set of inputs will arise.

THRESHOLD LOGIC IN THE QUR'AN

In systems theory, a *threshold* is a critical point at which small changes in input produce disproportionately large changes in output. Before this point, the system's behavior may remain relatively stable despite variations in input; beyond it, the system's state may shift abruptly, often irreversibly. In computational modeling, thresholds are represented through functions such as step functions – where an output "jumps" from one state to another – or sigmoid curves, where the rate of change accelerates rapidly once a tipping point is reached.

The Qur'an contains numerous examples of such threshold dynamics, where a single action, decision, or collective state can tip the moral and spiritual balance of individuals or societies.

- *A Single Act Tipping the Moral Balance*: In 5:32, the Qur'an declares that *"whoever kills a soul – unless in retribution for murder or for spreading corruption in the land – it is as if he had killed all of humanity; and whoever saves one – it is as if he had saved all of humanity."* This verse illustrates a moral threshold: the act of unjustly taking a single life is not merely an isolated transgression but one that carries the weight of an immeasurable collective wrong. Conversely, saving a single life is elevated to the level of saving all of humanity. In computational terms, the "output" here is not a proportional scaling of the act but a categorical jump from one moral state to another.
- *Crossing Between Belief and Disbelief:* The Qur'an often frames faith itself as a threshold condition. Crossing from belief (*iman*) to disbelief (*kufr*), or vice versa, is depicted as a decisive shift in spiritual state – one that reconfigures the entire relationship between the individual and divine guidance. The change may result from a single act of rejection or acceptance, but its consequences extend far beyond the immediate moment, affecting destiny in both this world and the next.
- *Societal Thresholds:* In 11:117, the Qur'an affirms that *"Your Lord would never destroy a society unjustly while its people were reformers."* This implies the existence of a societal threshold: as long as a critical mass of moral reform exists, societal collapse is averted. When that threshold is crossed – when collective sin and corruption surpass the balancing force of righteousness – the system shifts toward instability, often culminating in collapse.

From a CRS perspective, these threshold conditions can be formalized using mathematical models. Step functions capture the "all-or-nothing" transitions, while sigmoid functions can represent gradual build-ups that lead to rapid transformation once the tipping point is reached. Modeling such thresholds allows us to better understand the nonlinearity of divine systems: not all inputs lead to proportionate outputs, and some changes in moral or social state are both decisive and irreversible once enacted.

FUNCTIONS AND MAPPINGS

In mathematics and computational modeling, a *function* defines a specific mapping between an input set and an output set. Each input corresponds to one or more possible outputs, determined by the rules of the system. Within the Qur'anic framework, divine guidance can be seen as a network of such mappings, translating human and societal states into spiritual, moral, and material consequences.

At the most abstract level, Qur'anic guidance can be represented as:

$$f(Human_{State}) \longrightarrow Outcome_{State}$$

Here, $Human_{State}$ represents the moral, spiritual, and behavioral configuration of an individual or society at a given moment, while $Outcome_{State}$ represents the resulting consequence – whether divine reward, punishment, societal stability, or collapse. The Qur'an provides the *ruleset* for this mapping, with clear definitions, conditional clauses, and exceptions that refine how specific inputs are processed.

Examples of Qur'anic Functional Relationships:

- *Patience + Trust in God Victory:* In 3:200, believers are urged to *"Be patient, persevere, and remain stationed, and fear Allah so that you may be successful."* This creates a composite input – patience, perseverance, steadfastness, and God-consciousness – that maps to the output of success, often in the face of adversity.
- *Injustice + Arrogance Downfall:* The account of Pharaoh (28:39–42) illustrates a functional mapping in which persistent injustice and arrogance produce a systemic output of ruin, both materially and in legacy. The cause–effect mapping is not a simple moral observation but a systemic law encoded in revelation.
- *Charity Multiplication of Blessings:* In 2:261, charity is described as yielding returns *"like a grain that grows seven ears, each bearing a hundred grains."* This is a clear non-linear mapping – one act of giving leads to multiplied returns, suggesting a multiplicative function rather than a one-to-one proportional relationship.

From a computational standpoint, these relationships can be modeled in multiple ways:

- *Deterministic Functions:* When the Qur'an sets clear, unconditional cause–effect rules (e.g., certain punishments for specific crimes), these can be represented as deterministic mappings.
- *Probabilistic Functions:* When outcomes depend on multiple variables, including intention and context, probabilistic mappings better capture the uncertainty and complexity.
- *Composite Functions:* Some outputs result from the interaction of multiple conditions, requiring nested or chained functions to represent the full mapping accurately.

By formalizing these mappings mathematically, CRS enables a disciplined approach to understanding Qur'anic cause–effect logic. It becomes possible to simulate the implications of certain actions or societal states, explore the interplay of variables, and identify the points at which changes in input lead to significant shifts in output – linking this section back to the threshold logic just discussed.

Feedback Loops in Divine Systems

In both systems theory and cybernetics, *feedback* refers to the process by which a system's outputs are fed back as inputs, influencing future states of the system (Stacey, 2011). Feedback loops can be positive – amplifying the initial change – or negative – counteracting it to maintain stability. In divine systems, as reflected in the Qur'an, feedback operates at both the individual and societal levels, ensuring that guidance remains responsive, corrective, and adaptive over time.

Positive Feedback in the Qur'anic System:

Positive feedback reinforces virtuous action, creating a cycle that strengthens the moral and spiritual state of the believer. For example, in 2:261, acts of charity are said to multiply blessings many times over. These blessings – whether material, spiritual, or communal – encourage further giving, creating a self-reinforcing loop. Similarly, in 14:7, God promises: *"If you are grateful, I will surely increase you [in favor]."* Here, gratitude leads to increase, which fosters more gratitude, further amplifying the original input.

Negative Feedback in the Qur'anic System:

Negative feedback mechanisms serve as corrective forces, counteracting deviations from divine guidance. The Qur'an's warnings about sin, injustice, and corruption operate in this way. For instance, 7:96 explains that prosperity is granted when people believe and act righteously, but adversity and hardship may befall them when they persist in wrongdoing. These adversities are not merely punishments; they are corrective inputs intended to steer individuals or societies back toward equilibrium with divine law.

Cyclical Patterns and System Stability:

Across Qur'anic narratives, one can observe cycles: wrongdoing → divine warning → repentance → reform → stability. This cycle reflects a self-regulating system designed to prevent irreversible collapse as long as corrective feedback is heeded. However, the Qur'an also records cases where persistent rejection of feedback – where warnings are ignored – leads to system failure, as with past nations destroyed after exhausting all opportunities for reform (e.g., 11:116–117).

From a computational modeling perspective, these feedback loops can be represented as iterative functions where the output at time ttt becomes the input at time t+1t+1t+1. Positive feedback loops can be modeled using growth functions, while negative feedback can be represented through damping equations or control system models. In Complex Adaptive Systems (CAS), such loops are critical for adaptation: they enable the system to respond dynamically to changes in its environment, just as divine guidance adapts in application to new moral, social, and historical contexts.

In the Qur'anic framework, feedback loops ensure that guidance is not only a static rulebook but a living process – capable of reinforcing what is good, correcting what is wrong, and sustaining the long-term stability and vitality of the faith community.

Synthesis: Scripture as a Living, Adaptive System

In this chapter, we have moved from a foundational understanding of systems theory into a direct application of its principles to the Qur'an, revealing it as a dynamic, open, and complex adaptive system. Through this lens, scripture is not simply a historical record of divine communication; it is an operational architecture of guidance – continuously interacting with human actions, societal conditions, and the unfolding events of history.

We began by framing the Qur'an as an open system, receiving constant inputs from the lived realities of believers and generating outputs – beliefs, laws, ethical norms, and social structures – that reshape the environment from which new inputs will emerge. We then recognized its dynamic nature, where guidance remains eternal in principle yet adaptive in application, producing new insights for each era and context.

By identifying the Qur'an's Complex Adaptive System properties, we uncovered how emergence, self-organization, adaptation, and nonlinearity operate within divine guidance. This allowed us to pinpoint threshold conditions, where small changes in input – whether a single moral act or a collective shift in societal behavior – can trigger profound systemic transformation.

We explored functions and mappings as formal representations of how the Qur'an translates specific moral and spiritual states into concrete outcomes. These mappings range from deterministic rules to probabilistic relationships, often requiring composite functions to reflect the multi-variable reality of divine law in action.

Finally, we examined feedback loops – the cyclical processes of reinforcement and correction that allow divine systems to remain stable yet responsive. Positive feedback amplifies virtue, while negative feedback restrains corruption, with both operating as essential regulatory mechanisms in maintaining the health of the moral-spiritual system.

Together, these elements demonstrate that the Qur'an is not a static artifact but a living, interactive system. Its revealed concepts enter human civilization as dynamic specifications for moral, spiritual, and societal operation. These specifications can be formally modeled using mathematics and systems theory, allowing CRS to analyze guidance with rigor, precision, and predictive capacity – without diminishing its divine authority.

HOW DOES THIS FIT INTO THE AI REVOLUTION AND WHY STATIC KNOWLEDGE GRAPHS ARE NOT SUFFICIENT?

One way to challenge our project is to say: "Why bother turning the entire Qur'an into mathematics when we can already build an ontology and a knowledge graph from LLM-extracted triples?" That move is not trivial; a KG built from Qur'anic text can indeed do a lot of useful work. By extracting entities (prophets, communities, concepts like justice, mercy, hypocrisy, repentance) and relations (commands, prohibitions, causal links such as "leads_to" or "forbidden_for"), we can construct a graph in which nodes and edges encode "who is related to what." This structure supports better retrieval, thematic clustering, and a degree of symbolic reasoning.

Ontologies are Static and Can't Model Change

You can ask the system to find all verses connecting orphans and justice, or to infer that if the Qur'an condemns oppression and an action is a subtype of oppression, then that action is disfavored. For RAG-style systems and basic rule-based reasoning, this is already a genuine step forward. It gives you a reasoning layer in the sense of "if A then B" and "X is connected to Y," and it allows AI systems to navigate Qur'anic content far more systematically than keyword search or ad hoc citations.

But this remains a static, propositional layer, and that is where it falls short of what the Qur'an itself is doing. The Qur'an does not merely list propositions about good and evil, or record one-off stories; it models trajectories. It describes how individuals and nations move through time as belief and action interact. It depicts accumulation: repeated sins harden the heart; consistent charity purifies both wealth and self. It reveals feedback loops: injustice erodes trust; eroded trust destabilizes society; that instability invites further injustice and finally collapse. It contains thresholds: hearts can become sealed; nations can cross a point of no return after which destruction becomes inevitable. It articulates trade-offs: fear and hope in balance, justice and mercy in context, the interplay between individual responsibility and collective fate. A standard ontology and triple store flattens all this into simple edges such as (Oppression, leads_to, Downfall) or (Believers, commanded_to, Be_just). It cannot naturally represent how much oppression, over what period, under which structural conditions, or what happens if oppression declines while arrogance rises. It cannot express when heedlessness becomes a sealed heart, or how a community's trajectory changes when gratitude grows but inequality widens. Triples capture who is related to what; they do not capture how the system evolves and responds.

CRS Turns it into a dynamic universe

Converting the Qur'an into mathematical form is what turns revelation from a semantic network into a dynamic, simulatable architecture. Mathematics allows us to write down how key Qur'anic variables change over time. Instead of a bare statement like "injustice leads to downfall," we can model, for example, justice (J(t)) and social stability (S(t)) as time-dependent variables governed by equations such as

$$\frac{dJ(t)}{dt} = f(\text{law}(t), \text{wealth}(t), \text{power}(t))$$

and

$$\frac{dS(t)}{dt} = g\left(J\left(t\right), G\left(t\right), A\left(t\right)\right)$$

where (G(t)) denotes gratitude and (A(t)) denotes arrogance. Now the Qur'anic insight that injustice corrodes societies becomes a dynamic law: we can explore rates of change, non-linear effects, and tipping points. This is much closer to the scriptural language of increase and decrease, hardening and softening, rising and falling, than a static triple could ever be.

Degrees and Thresholds

Mathematics also gives us a natural language for degrees and thresholds. Triples are binary – either the relation exists or it doesn't – whereas the Qur'an talks in intensities: repeated disbelief, escalating hypocrisy, incremental guidance, and "atom's weight" granularity in deeds. We can define state spaces in which, for instance, belief (B) varies in ([0,1]), hypocrisy (H) becomes a function of misalignment between inner and outer states, and thresholds are encoded: once (A(t)) (arrogance) or (H(t)) cross certain critical values, the "sealed heart" regime switches on and the transition dynamics change qualitatively. That kind of modeling is essential if we want an AI system to reason not just that hypocrisy is "bad," but how hypocrisy grows, when it becomes catastrophic, and which patterns of repentance or reform plausibly reverse its trajectory.

Integrate with empirical data

Mathematical models also integrate far more naturally with empirical data than bare triples. A triple store can tell you that oppression is linked to divine displeasure, but it cannot be calibrated against historical or contemporary data in a systematic way. Once we have explicit variables and equations, we can begin to connect Qur'anic constructs to measurable proxies: corruption indices, inequality levels, social trust surveys, conflict incidence, institutional performance metrics. Parameters in the model can be estimated, adjusted, and stress-tested. That allows us to ask questions of the kind: "Given the Qur'anic structure of justice and oppression, how does this actual society look along those axes?" or "If we implement policy X to reduce a particular form of economic exploitation, what does the model predict for social cohesion over thirty years?" In that setting, religion is no longer merely commenting on secular models; it is shaping the structure of the models themselves. And from the AI side, most modern architectures – reinforcement learning agents, multi-agent simulations, learned world models – operate on continuous state spaces, reward functions, and transition dynamics. To design an RL agent whose reward function is constrained by Qur'anic justice, or a simulator whose environment updates according to Qur'anic moral dynamics, we need exactly the sort of mathematical objects we are building: value functions, transition kernels, differential or difference equations, not just symbolic edges.

Retains semantic purity

There is another, more subtle reason why ontology-only pipelines are insufficient: an LLM-extracted ontology is trained on general language use, not on Qur'anic epistemology. If we build our knowledge graph directly from LLM outputs, terms like “justice” will drift toward whatever global English usage suggests they mean; “‘abd” will be flattened to “slave” or “servant” in a sociological sense; “taw īd” will be reduced to “oneness” or “monotheism” as a category in comparative religion. In doing so, we quietly smuggle secular or extra-Qur’anic semantics into Qur’anic terms. Our approach with the Universal Quranic Variable Ontology and the mathematical layer reverses that direction. We begin by defining variables from the Qur’an and the classical sciences, grounding them in scriptural usage and authoritative commentary. We then embed those variables in equations and dynamics that reflect Qur’anic patterns of cause, effect, and moral consequence rather than Western legal or philosophical defaults. Only then do we build ontologies and knowledge graphs on top of that structure. Mathematics here is not just a tool for calculation; it is a way of anchoring meaning in revealed structure instead of corpus frequency. It ensures that when an AI system reasons about “‘adl” or “taw īd,” it is operating within a Qur’an-consistent architecture rather than unconsciously adopting the ambient semantics of the internet.

Can study and impact policy

One of the decisive advantages of a fully mathematical Qur’anic model over a pure triple-based ontology is its ability to integrate with empirical data. Triples are discrete and categorical; they tell us that oppression is linked to downfall, or that charity is linked to blessing, but they do not specify magnitudes, rates, or sensitivities that can be calibrated against the real world. A mathematical model, by contrast, defines explicit variables that can be tied to measurable proxies: levels of corruption can be approximated by corruption indices, inequality by Gini coefficients or wealth distributions, social trust by survey-based trust metrics, institutional justice by rule-of-law indicators, and so on. The model’s parameters – how strongly corruption erodes social cohesion, how rapidly gratitude stabilizes communities, what threshold of arrogance triggers destabilizing feedback – can then be estimated, adjusted, and stress-tested using historical and contemporary data. This allows us to ask structurally meaningful questions such as: “Given the Qur’anic structure of justice and oppression, how does this particular society score along those dimensions?” or “If we implement policy X to reduce a specific form of economic exploitation, which Qur’anic variables improve or degrade, and what does the model predict for social stability, belief, and moral health over the next thirty years?” At that point, religion is no longer merely commenting on social science and policy models from the outside; it is reshaping their internal architecture by contributing its own formally specified variables and dynamics.

Mathematical structure is native to modern AI

Another advantage is that mathematical structure is native to the way modern AI systems actually operate. A reasoning layer built only from triples is useful for symbolic checks and explanations – it can answer questions like "Is this proposed action consistent with obligation O?" or "We recommend option X because node Y in the Qur'anic graph is connected to node Z in this way." But reinforcement learning agents, multi-agent simulations, and world-model-based architectures do not reason in terms of discrete triples; they inhabit continuous state spaces, optimize reward functions, and evolve according to transition dynamics expressed as differential or difference equations (or learned analogues thereof). If we want an RL agent whose reward function is constrained by Qur'anic justice and mercy, we need justice, mercy, and oppression expressed as variables and reward-shaping terms, not just labels in a graph. If we want a societal simulator that evolves under Qur'anic moral dynamics, we need explicit update rules for belief, hypocrisy, gratitude, corruption, and social cohesion. If we want an LLM to "live inside" a Qur'an-aware world model, then its planning and evaluation layers must query a structured environment defined by Qur'anic transition kernels and value functions, not just retrieve verses connected by semantic edges. In all these cases, triples alone cannot supply what is needed; they do not yield reward functions, value functions, or transition kernels. Expressing revelation in mathematical form is therefore not a luxury – it is the necessary bridge that connects Qur'anic epistemology to the internal machinery of contemporary AI.

Putting all of this together, the relationship between triples and mathematics is not competitive but hierarchical. Ontologies and knowledge graphs are necessary; they give us the static skeleton of Qur'anic concepts and their connections, enabling retrieval and symbolic checks ("is this recommendation consistent with obligation O?"). But they are not sufficient. Full mathematical conversion animates that skeleton into a living, dynamic system that can be simulated, calibrated with data, and integrated into AI systems at the level where those systems actually function. A triple-based reasoning layer allows AI to talk about the Qur'an: to answer questions, quote verses, and trace thematic connections. A mathematical Qur'anic world model allows AI to live inside a Qur'anic environment: to evaluate policies, trajectories, and behaviors under the same dynamic logic that the text itself uses. That is the decisive gain, and it is why we insist on translating the entire Qur'an into a mathematically coherent system rather than stopping at an ontology extracted from LLM-generated triples.

REFERENCES

Cilliers, P. (1998). *Complexity and postmodernism: Understanding complex systems*. Routledge.

Gadamer, H.-G. (2004). *Truth and method* (2nd rev. ed.). (Weinsheimer, J., & Marshall, D. G., Trans.). Continuum. (Original work published 1960)

Holland, J. H. (1995). *Hidden order: How adaptation builds complexity*. Addison-Wesley.

Iser, W. (1978). *The act of reading: A theory of aesthetic response*. Johns Hopkins University Press.

Kauffman, S. A. (1993). *The origins of order: Self-organization and selection in evolution*. Oxford University Press. DOI: 10.1093/oso/9780195079517.001.0001

Meadows, D. H. (2008). *Thinking in systems: A primer*. Chelsea Green Publishing.

Mitchell, M. (2009). *Complexity: A guided tour*. Oxford University Press. DOI: 10.1093/oso/9780195124415.001.0001

Rahman, F. (2009). *Major themes of the Qur'an* (2nd ed.). University of Chicago Press.

Saeed, A. (2006). *Interpreting the Qur'an: Towards a contemporary approach*. Routledge., DOI: 10.4324/9780203016770

Stacey, R. D. (2011). *Strategic management and organisational dynamics: The challenge of complexity to ways of thinking about organisations* (6th ed.). Pearson Education Limited.

von Bertalanffy, L. (1968). *General system theory: Foundations, development, applications*. George Braziller.

Wiener, N. (1948). *Cybernetics: Or control and communication in the animal and the machine*. MIT Press.

Chapter 4
Introduction to Computational Quran

ABSTRACT

This chapter introduces the Computational Quran as a new epistemic modality that treats the Qur'an not only as a sacred text but as a structured, dynamic system capable of computational modeling. It argues that divine guidance is simple yet profound, inviting every generation to extract meaning anew rather than merely inheriting past interpretations. The chapter critiques historical limitations in tafsir, particularly its neglect of collective social processes, and shows how computational structuring can unlock the Qur'an's relevance for modern challenges, including AI, ethics, governance, and social sciences. By reframing revelation in mathematical and logical terms, the Computational Quran enables both humans and machines to engage its guidance, positioning scripture as an active force in shaping the emerging technological era.

INTRODUCTION TO COMPUTATIONAL QURAN

The Quran, the central scripture of Islam, is revealed to human societies by Allah (God). The Quranic term for this process is *nazool*, which does not refer to the physical descent of the Quran, such as an apple falling from a tree or a meteoroid landing on Earth. Instead, *nazool* signifies the transfer of divine knowledge to human cognition. The Quran itself declares that it is a *hidayat* (guidance) for humanity: "This is the Book about which there is no doubt, a guidance for those conscious of Allah" (*Al-Baqarah 2:2*).

These considerations about the Quran imply that God has created humans with the capacity to receive guidance through knowledge transfer. The essence of receiving this guidance is embedded in the very nature of human creation. Unlike other creatures, humans are endowed with the responsibility to seek perfection, and there-

DOI: 10.4018/979-8-3373-6940-2.ch004

fore, they continuously require divine guidance. The Quran, in its role as *hidayat*, serves as a source of this guidance, where Allah has chosen to share His knowledge with humanity. It informs humans about themselves, their purpose of existence, the principles that govern their lives, and the standards by which they should live. As Allah says in the Quran, "And We have certainly sent to every nation a messenger, saying, 'Worship Allah and avoid Taghut'" (*An-Nahl 16:36*).

This guidance is encoded within the Quran, and it requires human effort to decode and comprehend. While the Quran provides profound insights, it does not claim that its guidance is overly complex or demands extraordinary intellectual sophistication. In fact, the Quran explicitly asserts that it has been made easy to understand and accessible to all. Allah states in the Quran: "And We have certainly made the Quran easy to remember. But is there anyone who will be mindful?" (*Al-Qamar 54:40*). This indicates that the guidance within the Quran is inherently simple, but it is the responsibility of the human intellect and effort to engage with and extract its wisdom.

TAFSIR AND GENERATIONAL KNOWLEDGE TRANSFER

The process of extracting this guidance is known as *tafsir* (exegesis), which involves interpreting and explaining the meanings of the Quranic text. Tafseer is not merely a theoretical exercise; it is an active process where scholars and individuals seek to apply the Quranic knowledge to their lives. As Allah mentions: "Say, 'My Lord, increase me in knowledge'" (*Taha 20:114*), indicating that continuous learning and reflection upon divine revelation are fundamental to the human quest for understanding.

Therefore, the process of extracting guidance from the Quran is not time-constrained, nor is it limited to a single generation. It is not dependent upon the transfer of knowledge from one generation to the next. The Quran is considered timeless, and its timelessness necessitates that every generation actively seek guidance from it. As Allah affirms: "This Quran is not such as could ever be produced by other than Allah; but it is a confirmation of what was before it, and a clear explanation of the Book, wherein there is no doubt, from the Lord of the worlds" (*Yunus 10:37*). The Quran's guidance is universal, transcending time and context, and its wisdom remains relevant to all human circumstances.

The extraction of guidance from the Quran, therefore, is not dependent upon the interpretations or understanding of previous generations. Rather, it is a dynamic and ongoing process where each generation is encouraged to explore, interpret, and extract knowledge afresh, responding to the unique challenges and contexts of their time. The Quran itself emphasizes this perpetual engagement: "[This is] a blessed Book which We have revealed to you O Prophet so that they may contemplate its verses, and so that people of reason may be mindful." ād 38:29. This perpetual call to reflection ensures that the Quran remains a living, evolving source of guidance.

Furthermore, no generation can claim to have exhausted the depth of understanding offered by the Quran. As the Quran is intended to be the central guide for all of humanity, its miraculous nature requires that each generation continues to explore and extract knowledge from it, constantly engaging with its wisdom and applying it to new circumstances. This ongoing engagement with the Quran is itself a sign of its timelessness and divine wisdom.

NEEDS INTENTION AND SKILLS

While the Quran has been delivered to humankind and is always ready to provide knowledge, human societies must possess both the intention and the skill to extract its meaning. The Quran itself encourages active engagement with its text: "Say, 'My Lord, increase me in knowledge'" (*Taha 20:114*). This verse reflects the continuous, active effort required to seek understanding and wisdom from the divine revelation. Thus, every generation must be prepared to invest effort into uncovering the guidance embedded within the Quran, adapting its insights to their unique context.

One tradition that has emerged within Muslim societies advocates that the interpretations and guidance extracted by earlier generations should become the unquestioned standard for future generations. However, the Quran does not impose such a requirement. In fact, the Quran encourages independent reflection and thought. As Allah says, "And those who strive for Us – We will surely guide them to Our ways" (*Al-Ankabut 29:69*), emphasizing that intellectual and spiritual striving is not only permissible but encouraged.

This tradition of relying on the interpretations of earlier generations has, in some cases, hindered further research and exploration of the Quran. It has led to the acceptance of established interpretations as final and binding, discouraging independent scholarly inquiry and the application of new methodologies to Quranic exegesis. Modern generations, therefore, have been expected to follow these inherited interpretations, rather than conducting their own independent research and developing novel insights from the Quran. However, the Quran's timeless and living nature demands continuous engagement, ensuring that each generation brings fresh perspectives to its study.

While Muslims today deeply love, respect, and believe in the Quran, many fail to establish a profound intellectual relationship with it. This gap is a direct result of the belief that substantial and exhaustive work has already been completed by earlier scholars, leading to the perception that no further interpretation is either necessary or possible. The existing scholarly works, now housed in libraries and revered as authoritative, are often seen as the final word on the Quranic text. This mindset discourages contemporary scholars and individuals from exploring the Quran from new perspectives. As the Quran itself suggests, "Do they not reflect upon the Quran, or are there locks upon their hearts?" (*Muhammad 47:24*). This verse highlights the importance of continuous reflection and engagement with the divine message, emphasizing that intellectual engagement should not be stifled by the work of past generations.

The prevailing attitude towards Quranic studies has shifted the focus away from active research and exploration. Rather than viewing Quranic study as an ongoing journey of discovery and intellectual inquiry, it has become more of a path for consolidating and comparing the interpretations of the past. This shift positions Quranic scholarship not as a dynamic field of active exploration but as a static field of comparison. The effort to develop new tafsir (exegesis) perspectives is often seen as unnecessary, and is regarded as yielding diminishing returns. However, the Quran itself calls for a continual and evolving engagement: "Indeed, this Quran guides to that which is most suitable" (*Al-Isra 17:9*). The ongoing process of seeking guidance from the Quran is a dynamic one, and each generation is encouraged to bring new insights to the text.

POWER IN SIMPLICITY

The Quran is indeed easy to understand, but it possesses tremendous depth. The simplicity of the Quran does not imply that it lacks complexity or offers only surface-level concepts. On the contrary, the Quran's simplicity is an invitation to engage with its profound and layered meanings. As Allah declares: "Indeed, We have made the Quran easy to remember, so is there anyone who will be mindful?" (*Al-Qamar 54:40*). This verse underscores that the Quran is accessible, yet its depth requires thoughtful reflection and intellectual engagement. The Quran offers profound wisdom, and each verse contains multiple layers of meaning that can be unraveled through sincere effort and reflection.

The depth of the Quran makes it the responsibility of every generation to understand and interpret it in a manner that is both responsible and relevant to the times they are living in. Each era brings its own unique set of challenges, and it is incumbent upon scholars and thinkers to interpret the Quran in light of those challenges, ensuring its guidance remains applicable and impactful. As the Quran itself encourages, "Say, 'My Lord, increase me in knowledge'" (*Taha 20:114*), a call to continual intellectual and spiritual growth. This continuous engagement with the Quran is not only about understanding its historical context but also about applying its teachings to contemporary issues in a way that meets the evolving needs of society.

Hence, the goal of a *mufassir* (interpreter of the Quran) is not merely to convey the knowledge extracted by previous generations but to prepare their own generation to extract knowledge from the Quran that is relevant and transformative for their time. This task involves engaging with the depth and complexity of the present era, integrating the unique challenges and circumstances of the time into the interpretation of the Quran. As Allah commands: "And We have sent down to you the Book as clarification for all things" (*An-Nahl 16:89*), indicating that the Quran is a comprehensive guide for all periods, yet requires contextual understanding to address contemporary realities.

THE CHALLENGE OF MODERN TIMES

A *mufassir* must strive to bring out the guidance of the Quran in a manner that shapes responses to modern challenges, ensuring that its teachings are not confined to the past but are applied to the present and future. This dynamic process involves extracting principles from the Quran that can help address current issues, from social and political concerns to ethical and technological dilemmas. The Quran's timeless relevance demands that each generation actively engage in the interpretation of its verses, ensuring that the guidance it provides is both applicable and meaningful in the context of the modern world. As Allah says, "This is a book whose verses are perfected and then explained in detail" (*Hud 11:1*), highlighting the clarity and comprehensiveness of the Quranic message, while also emphasizing the need for thoughtful interpretation.

Throughout the ages, the Quran has been analyzed and explained both verbally, as reflected in speeches and talks (*khutbahs*), and in written form. A vast body of textual work has been produced by previous generations, including the interpretations of the Ahlulbayt, the Sahabah (companions of the Prophet), and numerous scholars from various schools of thought. This large collection of works highlights the diverse interpretations of the Quran across time and tradition. As Allah says, "And We have sent down to you the Book with the truth, confirming that which preceded it of the Scripture and as a criterion over it" (*Al-Imran 3:3*), indicating that the Quran serves as a standard to clarify and refine previous revelations, yet it remains open to diverse interpretations.

The diversity of these interpretations underscores an important point: ease of interpretation does not necessarily imply that all interpreters will arrive at the same or similar concepts. The Quran, despite its clarity, invites a range of perspectives. Interpretations can vary significantly across sects, within sects, and among different schools of thought. This diversity is a natural outcome of the Quran's depth and the varied contexts in which it is read. As the Quran itself acknowledges, "We have certainly made the Quran easy to remember, but is there anyone who will be mindful?" (*Al-Qamar 54:40*). The ease of comprehension does not negate the intellectual effort required to understand the layers of meaning, nor does it suggest uniformity in interpretation.

This diversity in interpretation reflects the richness of the Quran's message, which allows for multiple insights and viewpoints to emerge, enriching the broader Islamic tradition. It also emphasizes the importance of contextualizing the Quran's guidance according to the needs and challenges of each generation.

THE EVOLUTION OF EXEGESIS

In the early days of *tafsir* (Quranic exegesis), interpretations were largely based on *riwayaat* (traditions), which were grounded in the life and sayings of the Prophet Muhammad (PBUH). These traditions were traced to the explanations provided by those closest to the Prophet, including the companions (*Sahabah*), and also based on observations reported by those who recorded these narrations either directly or indirectly through transmitted sources. In this sense, each verse was explained using cross-referenced *ahadith* (sayings) and traditions, with a primary focus on the Prophet's life and his teachings. As the Quran mentions, "We did not send any messenger except to be obeyed by the will of Allah" (*An-Nisa 4:64*), highlighting the central role of the Prophet's teachings in guiding the understanding of the Quran.

In Sunni Islam, the sayings and actions of the *Sahabah* were used as primary support for these traditions. They were seen as the closest individuals to the Prophet, whose words and deeds were considered a critical source for interpreting the Quran. In Shia Islam, however, the words of the *Ahl al-Bayt* (the family of the Prophet) were regarded as firmly authoritative, based on their perceived direct or divinely guided connection to the Prophet. As the Quran mentions: "Say, 'I do not ask you for it any reward except for the love of [my] near relatives'" (*Ash-Shura 42:23*), which some interpret as supporting the central role of the Prophet's family in providing authoritative guidance.

THE EVOLUTION OF TAFSIR INQUIRY

As generations progressed, *tafsir* evolved to encompass a broader spectrum of inquiry, integrating philosophical, jurisprudential (*shar'i*), literary, and other intellectual dimensions. This allowed for a more multifaceted approach to understanding the Quran, reflecting the intellectual and social changes of the time. The Quran itself encourages a depth of thought, as seen in the verse: "Do they not reflect upon the

Quran, or are there locks upon their hearts?" (*Muhammad 47:24*), which calls for continued reflection and deeper engagement with the text.

In later generations, thematic interpretations also became an important aspect of *tafsir*. This approach analyzed verses from various chapters based on thematic and topical connections, allowing for a more holistic understanding of the Quran's message. Thematic *tafsir* not only enriched the interpretations but also provided a more structured approach to understanding the relationships between different Quranic themes. These various approaches – whether traditional, philosophical, or thematic – have all contributed valuable perspectives and created a wide spectrum of analytical frameworks for Quranic scholarship.

For the past several centuries, *tafsir* has also been used as a mechanism to defend the relative positions of various schools of thought and sects within Islam. In this context, *tafsir* became a vehicle through which different sects sought to justify their beliefs and theological or ideological foundations. Rather than purely aiming for an unbiased or universal understanding of the Quranic message, the interpretation of the Quran was often leveraged to affirm the truthfulness, veracity, or authenticity of a particular sect or school of thought. As Allah says, "If you disagree about anything, refer it to Allah and the Messenger" (*An-Nisa 4:59*), which underscores the ideal of returning to the divine source for resolution. However, historical practices have sometimes diverged from this ideal, with interpretations becoming intertwined with sectarian interests.

This tendency to use *tafsir* as a tool for sectarian defense led to the Quran being interpreted in ways that sought to validate specific doctrinal positions, often resulting in the reinforcement of ideological boundaries rather than the pursuit of collective understanding. Such practices have shaped Quranic scholarship in ways that sometimes obscure the Quran's universal guidance, turning it into a means for defending and perpetuating divisions. The Quran, however, calls for unity: "And hold firmly to the rope of Allah all together and do not become divided" (*Aali Imran 3:103*), encouraging Muslims to seek common ground rather than using interpretation as a means of division.

QURAN AND BROADER HUMAN LIFE

Despite the vast variety and depth of analysis in *tafsir*, there remains a significant lack of emphasis on developing Quranic exegesis for the collective processes of human life. For instance, social, economic, political, and other collective human processes have not been sufficiently explored through the lens of Quranic guidance. The Quran was sent as a guide for society as a whole, not just for individuals. From an Islamic perspective, the individual is not a separate entity from society but is an integral part of the broader social fabric. Allah says: "Indeed, this Quran guides to that which is most suitable" (*Al-Isra 17:9*), underscoring that its guidance is meant to shape the entirety of human existence, including both individual and collective aspects. Islam aims to establish a perfect society, and the Quran, as a mechanism of guidance, is referred to as *hudalil naas* – a guidance for all of humanity (*Al-Baqarah 2:185*).

This suggests that *tafsir* should not only focus on the individual aspects of Quranic guidance but also extend its wisdom to broader social, economic, and political processes. Yet, traditional *tafsir* has largely not addressed these larger human processes, which form what we today call the social sciences. This is not because the Quran does not offer guidance in these areas; rather, it is because *mufassirun* (interpreters) did not prioritize these dimensions of human life in their exegesis. The Quran provides comprehensive guidance for human society, covering aspects of justice, governance, economics, and social relations. As Allah says, "And We have sent down to you the Book as clarification for all things" (*An-Nahl 16:89*), implying that the Quran's scope is vast, encompassing all areas of human life.

The lack of focus on these collective processes in *tafsir* has created a significant void. Societies have, as a result, failed to integrate Quranic guidance into the structures of human civilization in a way that could shape social institutions, economic systems, and political governance. Quranic *tafsir* remained predominantly a philosophical, spiritual, and individualistic ethical system rather than a comprehensive social science movement. The Quran's social teachings were left under-explored, and its transformative potential in shaping human civilization was largely untapped. However, as the Quran itself affirms: "This is the Book about which there is no doubt, a guidance for those conscious of Allah" (*Al-Baqarah 2:2*). This holistic guidance, when fully realized, has the potential to shape every aspect of societal functioning, from governance to economy to interpersonal relations.

Not only did the desire or understanding of how to extract guidance from the Quran to build Quran-centered social sciences fail to manifest, but also the methods and frameworks for such a development did not evolve. As a side note, we must acknowledge that the works of scholars like Baqir al-Sadr and others in Islamic Economics represent targeted efforts in this area. However, these efforts did not evolve into a formal science and remained in the early stages of development. Despite this, significant strides were made in developing frameworks for social, political, and economic systems in the Western and Eastern worlds, which grew into scientific fields supported by mathematics and, more recently, advanced computational models. This divergence further increased the distance between the scientific development of these fields and Quranic exegesis, which had not yet been fully developed into a comprehensive system for social sciences.

As human society advanced, it increasingly appeared that Quran-centered social sciences were either impractical or impossible. The idea of analyzing human processes, including social, political, and economic systems, from the perspective of the Quran was gradually viewed as unpragmatic, undesirable, and even amusing. The concept of extracting practical and scientific knowledge from the Quran to guide human life in these domains became marginalized. This led to a situation where no serious scientific effort was directed toward developing social sciences in accordance with Quranic guidance.

The Quran itself, however, provides a clear framework for guiding all aspects of human society, stating: "And We have sent down to you the Book as a clarification for all things" (*An-Nahl 16:89*), a verse that affirms the Quran's comprehensive nature. Despite the lack of formalized frameworks, the Quran remains a vast source of wisdom that, if properly understood and applied, could inform the development of social sciences. Unfortunately, the perception that such an endeavor was impractical led to a significant gap between the Quranic teachings and the social sciences as they are practiced today. This gap not only hindered the development of Quran-centered social sciences but also contributed to a wider intellectual divide between the spiritual and scientific understanding of human societies.

THE NEED FOR COMPUTATIONAL QURAN IN THE AI ERA

The previous discussion leads us to the central theme of this book: the *Computational Quran*. Traditionally, scriptures have been approached through two primary modalities: the direct reading of the sacred text and its subsequent exegesis or commentary. These methods have served as the foundation for understanding the Quran throughout history. However, this book proposes and develops a third epistemic modality: the computational model. In this approach, the Quran is not merely read and interpreted but also structured mathematically – framing it as a dynamic system of interrelated variables, formal functions, and logical constructs. This transformation shifts the Quran from a static text to a dynamic model, making it not only accessible for human intellectual engagement but also intelligible to artificial intelligence systems, decision modeling, cognitive simulations, moral reasoning engines, and the broader universe of structured knowledge engineering.

This computational approach unlocks new possibilities for interacting with the Quran in ways that have not been explored traditionally, including the development of Quran-centered social sciences. The Quran itself encourages reflection upon its message: "Do they not then reflect upon the Qur'an? Had it been from other than Allah, they would have surely found in it many contradictions." Al-Nisā' 4:82 and "Do they not reflect upon the Quran, or are there locks upon their hearts?" (*Muhammad*

47:24). By applying computational modeling to the Quran, we not only facilitate deeper human engagement with its timeless wisdom but also lay the groundwork for integrating Quranic teachings into systems that address the complexities of modern human society, such as economic, political, and social dynamics. The idea of structuring the Quran as a dynamic, logical system offers a solid foundation for Quran-centered social sciences, where the teachings of the Quran can inform and shape decision-making frameworks and the evolution of human civilization.

Today, we are living through one of the greatest technological revolutions in human history. This era is not only defined by the internet, which serves as a common platform connecting all of humanity, but more importantly, by artificial intelligence (Naqvi & Munoz, 2020). AI is the force that is automating human work by creating machines capable of both physical and cognitive tasks. These machines accumulate experience, learn, and adapt. We are witnessing a total transformation in human life. The advent of AI will impact all fields of science, including the social sciences, and will undoubtedly shape the future of human society.

In the midst of this profound change, many have already written off religion, suggesting that it will no longer be a viable model for human guidance in the future. Some propose that machines themselves will create a new form of religion, and that the presence of intelligent machines in human society will inevitably alter human values. As new values and norms take shape, these critics argue, religion will have no place in the evolving social structures.

At this critical juncture, it is important to recognize that the computational representation of religion will not only offer religion a significant role in the AI revolution but will also prove the critics of religion wrong. The Quran itself is a source of wisdom that transcends time and circumstance, providing ethical and cognitive guidance that is just as relevant today as it was when first revealed. As Allah says: "Indeed, this Quran guides to that which is most suitable" (*Al-Isra 17:9*), indicating that its guidance remains adaptable and applicable to all generations, including the current one driven by AI.

For religion to participate meaningfully in the AI revolution and to guide society effectively, it must adopt an epistemological model that is most compatible with the AI era. This model, as we propose, is computational. By structuring religious teachings in computational terms – through logical constructs, variables, and dynamic systems – religion can actively shape the development of AI and human society in ways that resonate with the values and ethics it has long championed. This computational model will not only give religion a vital role in the future of society but also demonstrate that it has the intellectual and moral capacity to guide humanity in the age of machines.

WHAT IS COMPUTATIONAL QURAN?

The "Computational Quran" refers to the idea of structuring the Quran – not just as a sacred text of divine guidance but as a logical, dynamic system that can be understood, represented, and applied through computational methodologies. The term implies that the Quran is more than a traditional text to be read and interpreted; it is a rich, complex entity that can be modeled and analyzed computationally, much like any advanced scientific system.

a. The Quran as a Dynamic System

The Quran, as a dynamic system, means that it is not a static body of knowledge but a living, evolving set of interrelated principles, rules, and moral guidelines that can be mathematically modeled. The computational Quran conceptualizes the Quran as an intricate network of variables, relationships, and functions that define the natural, social, and ethical worlds. These interconnections allow for a continuous flow of meaning and application, adapting to new contexts and challenges while remaining deeply rooted in its original purpose: to guide human life.

b. Logical Constructs and Structured Knowledge

In a computational framework, the Quran's content can be broken down into logical constructs – entities, categories, rules, and functions – just as one would structure any system of knowledge in computer science, artificial intelligence, or mathematics. For example, ethical directives in the Quran could be encoded as rules or logical conditions governing human behavior. Similarly, moral and spiritual teachings can be represented as decision-making algorithms or as models that relate to human cognition and behavior. The Quran, in this sense, becomes a source of structured knowledge, one that is not confined to a purely theological or philosophical interpretation but is translatable into formalized models that can be used in modern technological applications.

c. Guiding Human Civilization Through Computation

The computational Quran provides a systematic framework for translating divine guidance into practical, actionable insights that can shape all facets of human society, including social, economic, political, and personal domains. By modeling Quranic teachings computationally, we can extract principles that guide decision-making, ethical reasoning, and governance. This modeling approach makes the Quran relevant to modern challenges, allowing for its teachings to inform and influence fields like AI, governance, economics, and even environmental sustainability. It bridges the sacred with the pragmatic, allowing for a harmonious integration of spiritual and worldly endeavors.

d. Interdisciplinary Engagement

The concept of the computational Quran necessitates an interdisciplinary engagement, bringing together traditional Quranic scholarship with modern fields like artificial intelligence, computational ethics, systems theory, and knowledge engineering. The Quran is not just a text of faith but also a repository of wisdom that can be examined through scientific and rational lenses. By developing models that align Quranic teachings with contemporary scientific and ethical frameworks, the computational Quran offers an innovative path for integrating religion into modern technological landscapes. It allows for the Quran to be studied and applied in ways that transcend traditional exegesis, offering new avenues for research and practical application.

e. Impact on Modern Social Sciences

The computational Quran also proposes the possibility of Quran-centered social sciences. Traditionally, social sciences have developed separate from religious frameworks, with their own models of human behavior, economics, politics, and social systems. However, by using computational models to analyze human processes through the lens of Quranic teachings, a new discipline can emerge – one that integrates religious wisdom with scientific inquiry. This could lead to new insights into human development, governance, economics, and social structures, grounded in the moral and ethical teachings of the Quran, and supported by computational methods.

f. AI and Moral Reasoning

One of the most exciting aspects of the computational Quran is its potential role in the development of AI and moral reasoning engines. As AI systems become more sophisticated, they will need ethical guidelines to navigate complex decision-making processes, especially in areas like healthcare, law, and autonomous systems. The computational Quran could provide a framework for moral reasoning, where AI systems are not just guided by human-designed ethical rules but also informed by divine moral principles. This could lead to the creation of ethical AI systems that are rooted in the moral imperatives outlined in the Quran, ensuring that machines act in ways that align with human dignity, justice, and fairness.

g. The Future of Knowledge and Faith

The computational Quran marks a new epoch in the intersection of knowledge and faith, where religious texts are no longer seen as antiquated or incompatible with the advancements of technology. Rather, the Quran's teachings can be understood as a timeless resource that evolves with human understanding and technological progress. The computational Quran invites us to reframe religious knowledge in a way that aligns with modern scientific inquiry, making it an integral part of future innovations in AI, cognitive science, and social governance.

THE PROMISE OF COMPUTATIONAL QURAN

The "Computational Quran" is a paradigm shift that transcends traditional interpretations of religious texts. It redefines the Quran as a dynamic system of structured knowledge, accessible not only to human scholars but also to artificial intelligence and computational models. This approach offers new possibilities for integrating Quranic guidance into modern scientific fields, particularly social sciences, while maintaining the Quran's central role as a guide for humanity. The computational Quran is not just an academic exercise; it is an essential framework for rethinking the relationship between faith and technology in the modern world.

As religion adopts a computational reasoning approach, it unleashes tremendous potential to shape human civilization on an equal footing with other social sciences. Traditional models, often based on secular, materialistic, or humanistic principles, cannot be directly compared to religion-based models. While secular models typically focus on empirical data and human-centered logic, a computational religious model integrates divine guidance with human experience, offering a profound ethical and moral framework that transcends the limitations of purely secular systems. This unique perspective enables the development of solutions grounded in both spiritual wisdom and practical application, providing a holistic view of human society.

By employing computational models to represent religious teachings, new approaches can be developed to address crises in human societies. These models allow for a more systematic, structured analysis of social, economic, and political issues, guided by principles derived from sacred texts. For example, in the face of global challenges such as poverty, climate change, or economic inequality, computational religious models can offer solutions that align with moral imperatives such as justice, equity, and compassion. These solutions are not merely theoretical; they are informed by a divine understanding of human dignity and the moral obligations of society.

Additionally, social justice and human rights can be viewed through a different lens when considered within the context of computational religion. While secular models often focus on human rights from a legal or political perspective, a computational religious model can introduce deeper dimensions, such as the inherent worth of every individual as part of a divine plan. It can also emphasize the interconnectedness of humanity and the moral duty to uphold the rights and dignity of others, not just as a legal obligation but as a spiritual one. The Quran itself emphasizes this concept of justice: "O you who have believed, stand firm for Allah, witnesses in justice..." (*Al-Ma'idah 5:8*). By integrating these ethical principles into computational frameworks, we can redefine how justice is understood and applied in the modern world.

In summary, the adoption of computational reasoning in religion not only strengthens its relevance in contemporary society but also positions it as a transformative force in the development of social sciences. By combining the divine wisdom of religious teachings with modern scientific methods, we can create more effective, equitable, and sustainable solutions to the challenges facing humanity today.

A NEW EPISTEMIC BEGINNING

The *Computational Quran* introduces a completely new way of understanding the Quran – not merely by reading the words or studying traditional interpretations, but by treating it as a structured system that can be analyzed, modeled, and even understood by intelligent machines. Traditionally, the Quran has been studied for its guidance, stories, and spiritual teachings. For centuries, people have relied on reading, interpreting, and reflecting upon the text to derive its wisdom. However, in today's world – shaped by advances in computers, data, and artificial intelligence – we now have the tools to take a significant leap forward: translating the Quran's wisdom into logical models, systems of relationships, and mathematical structures that can be understood not only by humans but also by machines.

This book demonstrates how the Quran can be expressed in the language of structure – using carefully defined terms, models, and equations – similar to how engineers design systems or doctors model the human body. This approach doesn't replace traditional religious study; rather, it adds a new lens, one that allows the Quran to speak in a format that the future can understand – the language of data, systems, and artificial intelligence. By using this computational model, we open up new possibilities for how the Quran's teachings can be applied, particularly in our increasingly technological world.

Just as we now have computers that can understand human language, music, and images, this book paves the way for computers – and scholars alike – to understand the spiritual and ethical meanings within the Quran in a structured and measurable way. This opens the door for the Quran to guide emerging technologies – ranging from education to ethics to decision-making – and invites both believers and nonbelievers to approach scripture with fresh eyes. The *Computational Quran* is not merely a reinterpretation; it is a revolutionary perspective that bridges timeless spiritual wisdom with the intellectual and technological landscape of the future.

Why is it unique?

This is the first book of its kind, and it is part of a two-part series. The second volume, *Computational Bible*, will apply the same computational approach to the Old and New Testaments, exploring how ancient scriptures can inform modern technologies in similar ways.

To our knowledge, this book is the first to apply principles from systems theory, mathematics, logic, and artificial intelligence to scripture as a formal body of knowledge. It pioneers a new approach by treating the Quran not just as a sacred text to be interpreted, but as a dynamic, structured system that can be analyzed, modeled, and applied through computational methods. While there are a few categories of books that may appear tangentially related, this work stands apart in its foundational shift toward epistemic modeling and computational structuring of the Quranic revelation.

a) *Traditional Quranic Exegesis and Modern Commentary:* Books in this category, such as *The Study Quran* (Nasr et al.,2015) and thematic interpretations by scholars like Fazlur Rahman's *Major Themes of the Qur'an* (Rahman, 2009), offer valuable insights into the Quran through interpretive traditions and textual analysis. However, they remain firmly within the domain of human exegesis. These works do not employ formal modeling or present the Quran as a structurally computable system. Instead, they focus on contextual, historical, and philosophical interpretations – valuable as they are, they do not engage with the Quran in a computational framework.
b) *Digital Humanities and Religion:* Books in this category explore how religious texts are digitized, indexed, or analyzed using computational linguistics. Examples include *Digital Humanities and Islamic & Middle East Studies* or *Computational Analysis of Religious Texts* (Muhanna, 2016). While these works incorporate technology to manage and analyze textual data, they focus on data science and textual processing rather than epistemic modeling or mathematical structuring of divine revelation. These books do not aim to render religious content computable or systematically integrate it into intelligent systems.
c) AI and Ethics / Religion and Technology: A growing number of books address the intersection of AI and theology, such as *God, Human, Animal, Machine* by O'Gieblyn (2021) or *Religion and AI* (Singler, 2024) by multiple academic authors. These texts are primarily philosophical reflections on how AI impacts religion and ethics, but they do not offer computational implementations or models.

They explore the implications of AI on faith and theology without attempting to make religious content usable within intelligent systems or computational environments.

d) *Books on Quran and Science:* Popular works like *The Bible, the Qur'an, and Science* by Bucaille (2003) attempt to align scientific knowledge with Quranic verses. These works, while valuable in their own right, often focus on apologetic or polemical arguments, aiming to prove the Quran's scientific accuracy. They do not provide rigorous systems modeling or frameworks for AI-based analysis, and their focus is more on defending the Quran rather than structuring it for computational engagement.

We believe *Computational Quran* occupies a completely new category within the landscape of Quranic literature. It is not a commentary, not an apologetic explanation, not a digital catalog, and not a philosophical essay on AI and faith. Rather, it is a field-defining work – the first to treat the Quran as a formal system of variables, functions, and ethical logic, ready to be studied computationally and integrated into intelligent systems. This work is both non-competitive and complementary to existing literature. It introduces a novel paradigm and may serve as a foundational reference point for an entirely new wave of comparative, computational, and faith-integrated scholarship that brings religious texts into alignment with the future of technological innovation.

ISLAMIC THEOLOGY (KALAM) AND ITS EVOLUTION

Islamic theology, known as *kalam* – meaning "speech" or "discourse" – is a sophisticated intellectual tradition that has developed over more than a millennium. It has been shaped not only by internal debates within the Muslim community but also by sustained engagement with other religious and philosophical traditions, including Christian theology and Hellenistic philosophy. Throughout its history, *kalam* has functioned as a dynamic arena for negotiating the relationship between revelation, reason, and lived experience.

Early Debates (7th–8th centuries)

In its earliest phase, Islamic theology was not a fully systematized discipline but rather a response to urgent political, ethical, and doctrinal questions arising after the Prophet Muhammad's death.

- **Succession and Leadership.** Disputes over who should lead the Muslim community – rooted in questions of legitimate authority – contributed to the Sunni–Shi'a divide, which continues to carry theological implications.
- **Faith and Sin.** The moral status of the grave sinner sparked controversy. The Kharijites argued that committing a major sin rendered a person no longer a believer, while the Murji'ites maintained that ultimate judgment belonged to God alone, leaving the sinner's faith intact.
- **Free Will vs. Predestination.** How to reconcile God's absolute sovereignty with human moral responsibility became an enduring theological problem, setting the stage for centuries of debate.

The Impact of Greek Philosophy (8th–10th centuries)

During the Abbasid Caliphate, the translation movement brought the works of Aristotle, Plato, and the Neoplatonists into Arabic, profoundly influencing Islamic intellectual life.

- **The Mu'tazilite School.** Often described as the "people of God's unity and justice," the Mu'tazilites pioneered systematic *kalam* using rationalist methods and Greek logic to defend Islamic belief. They argued that God's attributes were inseparable from His essence, that the Qur'an was created rather than eternal, and that human free will was essential to divine justice.
- **The Ash'arite and Maturidi Schools.** Reacting to what they saw as the Mu'tazilites' overreliance on reason, these schools sought a middle path between pure rationalism and literalism. The Ash'arites, founded by Abu al-Hasan al-Ash'ari, affirmed that God's attributes were real and distinct from His essence and that the Qur'an was uncreated. Their doctrine of *kasb* ("acquisition") held that God creates all human actions, but humans "acquire" them, preserving a form of moral accountability. The Maturidis, led by Abu Mansur al-Maturidi, shared much with the Ash'arites but placed greater emphasis on human reason as a means to know God and discern certain moral truths. Together, these schools became the dominant theological frameworks of Sunni Islam.

Synthesis, Systematization, and the Rise of Sufi Theology (11th century onwards)

The classical period saw the maturation of Sunni theology into fully articulated systems.

- **Doctrinal Systematization.** Scholars expanded and refined Ash'arite and Maturidi frameworks, producing detailed creeds (*aqidah*) and comprehensive theological treatises that provided rational proofs for foundational beliefs.
- **Sufi Theology.** Parallel to these rational-theological developments, Sufism (Islamic mysticism) offered an experiential approach to theology grounded in personal spiritual transformation and direct knowledge (*ma'rifah*) of God. Sufis explored the inner meanings of worship, the dynamics of divine love, and the stages of the soul's journey toward God. While sometimes in tension with the dialectical *kalam* tradition, Sufi thought often enriched theological discourse, offering dimensions of meaning inaccessible to purely rational argumentation.

Modern and Contemporary Theology

In the modern period, *kalam* has confronted new intellectual and political challenges.

- **Colonialism and Modernity.** Encounters with European colonial powers and the rise of Western modernity prompted Muslim intellectuals to grapple with questions of political sovereignty, cultural decline, and religious reform. Thinkers such as Muhammad Abduh and Jamal al-Din al-Afghani sought to reinterpret Islamic teachings for a modern context without abandoning foundational principles.
- **Science and Secularism.** Contemporary debates include the compatibility of Islamic theology with evolutionary theory, the role of reason in interpreting revelation, and the place of religion in secular or pluralistic societies. These conversations often echo earlier *kalam* debates, but now occur in dialogue with modern science, global ethics, and new philosophical currents.

How Islamic Theology Is Studied

The study of Islamic theology (*kalam*) has historically been rooted in religious institutions dedicated to the preservation, transmission, and defense of the faith. In the modern era, it has also become a subject of academic inquiry, where it is approached from a broader, often secular, perspective. While the aims and methods differ, both approaches engage with the intellectual richness and historical complexity of Islamic theology.

Traditional Method

In the traditional setting – whether in madrasas, scholarly circles (*halaqat*), or specialized institutes – the study of *kalam* is immersive and integrative.

- **Textual Mastery.** Students undertake deep study and, often, memorization of the Qur'an and Hadith (the sayings, actions, and tacit approvals of the Prophet Muhammad), which form the primary sources of Islamic theology.
- **Engagement with Classical Texts.** A central component is the study of classical theological works from major schools, including the writings of figures such as al-Ash'ari, al-Maturidi, and al-Ghazali. These works are approached not merely as historical documents but as living repositories of theological reasoning.
- **Logic and Debate.** The pedagogy is often dialectical, training students in formal logic (*mantiq*) and structured debate to defend core tenets of faith and refute opposing or heterodox positions.
- **Interconnection with Other Sciences.** Theological training is closely integrated with other Islamic sciences – *fiqh* (jurisprudence), *tafsir* (Qur'anic exegesis), and Hadith studies – reflecting the holistic nature of traditional Islamic scholarship.

Academic and Modern Method.

In contemporary universities, Islamic theology is often studied as a sub-discipline of Islamic studies or religious studies, with a methodological orientation that is typically more descriptive and analytical than confessional.

- **Historical and Contextual Analysis.** Scholars examine how theological ideas have developed in specific historical, political, and social contexts, tracing their evolution over time.
- **Comparative Approach.** Islamic theology is often studied alongside other theological traditions, such as Christian and Jewish thought, to illuminate similarities, differences, and points of intellectual exchange.
- **Interdisciplinary Tools.** Modern researchers draw from philosophy, intellectual history, anthropology, and the broader field of religious studies to analyze *kalam* texts and traditions, situating them within global intellectual history.

Islamic theology is thus a dynamic field of inquiry, with its own internal logic, core commitments, and evolving trajectories. The traditional approach prioritizes the faithful preservation and application of theological principles, while the academic approach seeks to understand *kalam* as a multifaceted intellectual and social phenomenon. Together, they reflect the dual nature of theology as both a living tradition and a subject of critical, historical study.

From the perspective of Computational Religious Studies, these two modes of engagement are not competing but complementary. CRS provides a means to model theological systems in a way that is faithful to their internal reasoning while also making them analyzable in broader academic and cross-disciplinary contexts.

REFERENCES

Bucaille, M. (2003). *The Bible, the Qur'an and science: The Holy Scriptures examined in the light of modern knowledge* (Pannell, A. D., Trans.). Tahrike Tarsile Qur'an.

Muhanna, E. (Ed.). (2016). *The digital humanities and Islamic & Middle East studies*. De Gruyter., DOI: 10.1515/9783110376517

Naqvi, A., & Munoz, M. (2020, March). *Rebuilding a company with artificial intelligence*. California Management Review. https://cmr.berkeley.edu/2020/03/rebuilding-a-company-with-ai/

Naqvi, J. (2022–2025). *Various lectures on YouTube* [Video series]. YouTube.

Nasr, S. H., Dagli, C. K., Dakake, M. M., Lumbard, J. E. B., & Rustom, M. (Eds.). (2015). *The Study Quran: A new translation and commentary*. HarperOne.

O'Gieblyn, M. (2021). *God, human, animal, machine: Technology, metaphor, and the search for meaning*. Doubleday.

Rahman, F. (2009). *Major themes of the Qur'an* (2nd ed.). University of Chicago Press.

Singler, B. (2024). *Religion and artificial intelligence: An introduction*. Routledge., DOI: 10.4324/9781003256113

Section 2

Methodology and Mathematical Framework

Chapter 5
The Five-Step Modeling Framework

ABSTRACT

Chapter 5 introduces the Five-Step Modeling Framework, the methodological backbone of Computational Religious Studies (CRS), designed to translate Quranic verses into computationally analyzable models without compromising textual sanctity or exegetical integrity. The chapter details a disciplined pipeline—canonical source text, tafsīr-grounded semantic summary, standardized variable mapping via UQVO, mathematical and formal expression, and thematic integration into a global scriptural map. Each step is shown to be essential, ensuring traceability, reproducibility, and structural coherence across verses. By preventing both reductionism and vagueness, the framework enables scalable, auditable, and AI-ready modeling of revelation, while remaining scripture-agnostic and adaptable to other religious corpora.

INTRODUCTION

The goal of this chapter is to fully delve into the methodology for building a computational model of a Quranic verse. We will introduce and analyze the Five-Step Modeling Framework that has been developed, as well as understand why this particular approach is used and why shortcuts are not acceptable. Each step will be explained in detail, showing how spiritual and moral meaning is translated into a structured, computationally analyzable form. This framework becomes the foundation and backbone of the field of Computational Religious Studies (CRS), and everything that follows in this book assumes and builds upon it.

The fundamental goal of building this modeling framework was to ensure that meaning is not lost from the original work, while still having created a sufficient enough model that the work can be computationally analyzed, simulated, and reasoned over. When we approach religious texts, we must also safeguard that the

DOI: 10.4018/979-8-3373-6940-2.ch005

original sanctity of the work is not lost, diluted, or distorted during the modeling process. The framework exists to keep that boundary clear while making the internal structure of guidance visible.

To align with these requirements, the following five-step method was developed:

- Source Text – Canonically precise extraction and formatting of the verse(s).
- Tafsīr (Exegesis) Summary – Condensed, tafsīr-grounded semantic understanding.
- Variable Mapping & Modeling – Mapping meaning into a stable variable system (UQVO) and conceptual model.
- Mathematical/Formal Expression – Expressing that model as equations, logical rules, graphs, or dynamic systems.
- Thematic Labeling & Integration – Tagging and positioning the model within a global, scripture-wide map.

Step 1 preserves the divine wording. Steps 2 and 3 interpret it. Step 4 formalizes it into a computationally usable structure. Step 5 connects each local model into a larger computational atlas of the Revelation. When viewed as a pipeline, this process ensures that the text is never treated as random data but as structured guidance whose organization can be made explicit, without stepping outside the bounds of accepted interpretation.

Although this method is presented here in the context of the Quran, the framework itself is not Quran-specific. The principles – preserving canonical text, grounding meaning in recognized exegesis, standardizing variables, formalizing structure, and integrating models into a global map – can be applied to other religious scriptures and corpora. In that sense, the Five-Step Modeling Framework is not merely a tool for one book; it is a blueprint for the entire field of CRS.

Why a Formal Framework Is Necessary

It is worth pausing to ask why such a structured framework is needed at all. After all, religious texts have been studied for centuries without recourse to variable ontologies or dynamic systems. Classical tafsīr, theology, law, and spirituality have already mapped many of the relationships between belief, practice, and outcome. Why introduce a computational layer?

The answer is not that traditional methods are lacking in depth. On the contrary, they are extremely rich. The challenge is that they are not structured for computation. For a human scholar, a thousand-page tafsīr can be navigated through intuition, memory, and years of training. For a machine, that same tafsīr appears as unstructured text unless we deliberately shape it into something more formal – variables, relationships, constraints, and systems.

Without a careful framework, any attempt to "model" scripture risks one of two extremes. On one side is reductionism, where divine speech is flattened into simplistic rules or numerical scores. On the other side is vagueness, where the language of "systems," "dynamics," or "emergence" is invoked without actual structure behind it. The Five-Step Framework exists precisely to navigate between those extremes, giving us a disciplined way to move from Revelation to model. This framework also guards against another subtle danger: arbitrary or idiosyncratic modeling. Without a standard pipeline, each verse could be modeled differently depending on who is doing the work, which would make cross-verse or cross-surah analysis incoherent. By enforcing a common sequence – text, tafsīr, variables, math, integration – we ensure that every modeled verse passes through the same stages of scrutiny and structuring. That repeatability is what makes CRS a science, not just an interesting intellectual exercise.

OVERVIEW OF THE FIVE STEPS AS A PIPELINE

Although we will go into each step in detail, it helps to view them as stages in a single flow. Step 1 gives us the anchor: the exact text. Step 2 gives us the semantic content as understood by scholars. Step 3 converts that content into variables and conceptual structures – the building blocks of a model. Step 4 transforms those structures into formal expressions – functions, logical rules, graphs, dynamic systems. Step 5 then locates that formal model within a global map, connecting it with other verses and themes.

The key idea is that no step is optional. If we skip Step 1, we lose the anchor. If we skip Step 2, we lose guidance from tafsīr and drift into personal speculation. If we skip Step 3, we have no standardized language for the computer to reason with. If we skip Step 4, we have descriptions but no formal models. And if we skip Step 5, we end up with isolated islands of analysis that never form a coherent atlas of guidance.

With that in mind, we can now walk through each step in depth.

Step One: Source Text

The first step is deceptively simple: extract the verse or verses to be modeled in a canonical form. In practice, however, this step carries enormous weight. It seems trivial until you try to model at scale and realize that every inconsistency in the source layer multiplies downstream.

For the Quran, Step One usually includes the following elements. First, the Arabic text of the verse is taken from a canonical edition. This text is treated as immutable. Second, a designated English translation is selected to serve as the primary language of modeling, not because it replaces the Arabic, but because the subsequent mathematical and conceptual structures will be described in a single working language for consistency. Third, precise metadata is attached: surah number, verse number, textual boundaries, and any segmentation decisions (for example, when multiple verses form one conceptual unit).

In the context of CRS, the Source Text is best thought of as a data object rather than a paragraph. It has fields: Arabic text, translation, identifier, and links. It is never edited, "fixed," or paraphrased in the modeling pipeline. If a translation is updated or refined, that becomes an explicit version change, not a silent modification. This discipline is critical. A computational system can only be as reliable as its anchors, and in this framework, the anchors are the verse-objects created in Step One.

This step also encodes a theological stance: that the modeled verses are not raw material to be manipulated at will but are the fixed reference points around which all interpretation and modeling must orbit. The entire framework is structured so that at any time, a modeler, scholar, or auditor can trace a variable, a function, or a graph node all the way back to the exact wording of revelation from which it was derived. That traceability begins here.

It is also in Step One that we decide on the granularity of modeling. Sometimes a single verse can be modeled as one unit. At other times, several verses form an inseparable conceptual block (for example, a sequence describing a law, a promise, or a narrative scene). The framework is flexible on whether the modeling unit is one verse or multiple, but it is strict about documenting that decision. Once a "modeling unit" is defined, that unit is treated consistently through the remaining steps.

Thus, Step One is not an afterthought. It is where we declare, in an explicit and structured way, what exactly we are modeling. Everything else depends on this.

Step Two: Tafsīr Summary

Once the text is anchored, we move to meaning – but not any meaning. Step Two is where we bring in tafsīr, the long-standing tradition of exegesis, and compress it into a structured, manageable semantic summary.

The purpose here is not to produce a complete commentary. Instead, we are trying to extract the elements of tafsīr that are structurally relevant to modeling. We are looking for relationships, conditions, roles, and implications that can later be mapped to variables and equations. Tafsīr becomes the raw material for structural insight.

This process usually involves consulting multiple tafsīr sources – classical, early, and sometimes contemporary – while paying attention to areas of consensus versus areas of difference. The goal is not to flatten disagreement, but to avoid anchoring a computational model in highly controversial or niche readings unless that plurality itself is part of what is being modeled. Major points of agreement are highlighted as foundational. Conditions, exceptions, and nuances are captured where they affect structure: for example, when a promise is tied to specific preconditions or when a legal ruling depends on context.

A key challenge in Step Two is semantic compression. Tafsīr can be vast. A single verse may have dozens of pages of commentary. A computational model, however, cannot afford to carry narrative detail at that scale. Instead, we extract a set of concise semantic statements, such as:

- This verse establishes a conditional relationship between X and Y.
- This verse defines a category of people characterized by states A and B.
- This verse links a set of behaviors to an outcome, in this life or the next.
- This verse describes a feedback loop between internal states and external actions.

The result is a small number of carefully written sentences that preserve the structural content of tafsīr without drowning the later steps in redundancy. These statements are still human-readable; they are not yet variables or equations. But they are shaped with Step Three in mind. You can almost see the variables and arrows hiding inside the sentences.

Another important dimension of Step Two is epistemic humility. The modeler must recognize that their role is to curate meaning from tafsīr, not to invent it. When the tradition is unclear or divided, that uncertainty should be captured in the summary rather than overwritten by private judgment. Computational models are powerful; if they are built on speculative readings presented as if they were certainties, they will mislead both machines and humans. Step Two is where we decide, explicitly, that we will not do that.

Step Three: Variable Mapping & Modeling

Step Three is the pivot point between human-centered exegesis and machine-usable structure. Here, we take the semantic statements from Step Two and convert them into a system of variables and relationships built on the Universal Quranic

Variable Ontology (UQVO) – the standardized set of Quranic variables introduced later in this book.

In practical terms, this means identifying all of the conceptual elements in the tafsīr summary and assigning them to variables from a shared library. Internal states like belief, hypocrisy, fear, hope, or arrogance are mapped to cognitive and spiritual variables. Actions like prayer, charity, oppression, or lying are mapped to behavioral variables. Social conditions like justice, corruption, cohesion, or civil strife become social variables. Outcomes like guidance, misguidance, blessing, punishment, or societal collapse are mapped to outcome variables. Temporal dimensions are represented as trajectory or time-indexed variables.

The key here is standardization. If a verse speaks about belief and righteous action, the variables it uses for those concepts should be the same as those used in every other verse that deals with them. This is what allows cross-verse analysis. It also forces a discipline: we cannot casually rename variables or invent new meanings just to fit a local verse. Instead, we must align that verse with a stable ontology and, if necessary, refine the ontology in light of what the verse reveals.

Once the variables are chosen, Step Three continues with conceptual modeling. We ask questions like:

- Which variables are inputs? Which are outputs?
- Which variables represent conditions or pre-conditions?
- Are there clear causal links (A leads to B)?
- Are there feedback cycles? For example, does belief lead to action which then reinforces belief?
- Are the relationships conjunctive ("and"), disjunctive ("or"), negating ("not"), or threshold-based?

The answers are first expressed in plain language, then organized into conceptual diagrams – often implicitly rather than as literal drawings: "Belief and righteous deeds together place an individual into a class of those promised a specific outcome." That sentence already encodes structure: a joint condition, a defined class, and a linked outcome.

Step Three therefore produces two things: a set of variables, each linked back to UQVO, and a conceptual architecture describing how those variables relate. It is at this point that the verse is no longer just words and commentary; it has become a structured system whose components can be manipulated, analyzed, and compared.

Step Four: Mathematical and Formal Expression

If Step Three is where we create the conceptual skeleton, Step Four is where we give it a mathematically articulated body. The relationships defined in Step Three are now expressed using formal tools: functions, logical rules, probability structures, graph theory, or dynamic systems. The choice of tools depends on the nature of the verse, but the guiding principle is always the same: to express the structure as clearly as possible. For example, a verse's tafsīr that says, in effect, "Those who believe and do righteous deeds are promised Paradise," can be expressed as a logical rule with thresholds and a binary outcome: if belief exceeds a certain level and righteous action exceeds a certain level, then the individual belongs to the class of those granted this promise. That rule can be written as an equation, as a piecewise function, or as a logical implication.

Other verses may describe processes rather than simple conditions: the gradual hardening of hearts through repeated sin, the improvement of character through repeated acts of charity, the long-term destiny of a community depending on its collective behavior. These lend themselves to dynamic modeling using difference equations or differential equations, where a variable like belief or social cohesion changes over time based on the current level of obedience, injustice, remembrance, or arrogance. In these cases, the verse is captured not by a single condition, but by an update rule that can be iterated or simulated.

Graph theory is another tool frequently used in this step. Variables become nodes in a graph, and their relationships – causal, conditional, reinforcing, inhibiting – become edges. This allows us to see, within a single verse and across verses, which concepts are central, which ones mediate between others, and how tightly the moral and spiritual universe of the Quran is interconnected. It also sets the stage for more advanced network analyses later on.

A crucial part of Step Four is remembering what we are not doing. We are not assigning "scores" to souls or pretending we can numerically quantify divine mercy and justice. What we are doing is modeling the *structure* of the guidance that the verse provides: how beliefs, actions, states, and outcomes are logically and behaviorally related. Mathematics, in this frame, is not a tool of reduction but a language of clarity.

Another benefit of formalization is that it makes the model executable. Once written in equations or logic, a verse-based model can be integrated into an AI system or a simulator. This allows us to explore "what-if" scenarios, to examine the emergent behavior of interconnected models, and to study how certain patterns of belief and action might propagate through individuals or societies. None of this replaces the spiritual or ethical responsibility of the human, but it does open a new layer of analysis: the study of Revelation as a system.

Step Five: Thematic Labeling and Integration

If we stopped at Step Four, we would have a series of beautifully constructed but isolated models – each faithful to its verse, each mathematically clear, but not yet connected to the rest of the Revelation. Step Five is where those isolated islands are turned into a continent.

This step has two main components: thematic labeling and global integration. Thematic labeling assigns each model to one or more conceptual categories. These categories reflect major domains of Quranic discourse: faith and spiritual cognition, worship and devotion, family and social roles, economic justice, law and ethics, war and peace, leadership and governance, history and eschatology, and so on. Within each major domain, there can be more specific sub-themes.

By tagging each modeled verse with these themes, we create a way to retrieve and analyze them in groups. For example, we can pull all verses whose models involve charity as a variable and see how the Quran treats charity across contexts – individual, communal, legal, spiritual. Or we can study all verses related to hypocrisy and see how that concept is structured over the entire corpus.

Global integration goes a step further. Because all verses are modeled using the same variable ontology (UQVO) and expressed in compatible formal languages, we can literally connect their models into a single knowledge graph or system. Belief in one verse is the same variable as belief in another. Justice in one legal context is directly linkable to justice in a narrative or eschatological context. Step Five recognizes those overlaps, aligns them, and merges the local graphs into an overarching network.

This is where the Quran reveals its systemic nature most clearly. What might look, on a page, like separate topics scattered across surahs becomes, in the integrated model, a tightly coupled epistemic structure: beliefs leading to actions, shaping individuals, forming communities, affecting history, and determining ultimate outcomes. Step Five is where CRS stops being a collection of models and becomes a map of Revelation.

Another important function of Step Five is enabling comparative work. Because each verse-model is thematically tagged and structurally integrated, we can later place models from other scriptures or traditions into analogous positions and study how they align or differ. The framework does not demand that all scriptures say the same thing; it simply gives us a disciplined way to compare their structures. That comparative potential depends heavily on the integrity of Step Five.

HOW THE FRAMEWORK BECOMES THE BACKBONE OF CRS

Having walked through the five steps, it should now be clearer why this framework is more than just a helpful workflow. It is the backbone of Computational Religious Studies, for several reasons.

First, it enforces methodological discipline. CRS lives at the intersection of theology, computer science, and mathematics. Each of those disciplines has its own internal standards. The Five-Step Framework functions as a common contract across them: theologians and scholars can trust that meaning is grounded in tafsīr and anchored in the text; mathematicians can trust that the structures are rigorous and consistent; computer scientists can trust that the models are explicit and machine-readable. Without a shared framework, CRS would risk becoming a loose cluster of individual projects with no coherent standards.

Second, it enables scalability. A single verse can be modeled in many ways. What enables modeling at the scale of a chapter, a surah, or the entire Quran is not just effort, but standardization. Because every verse passes through the same five steps, the work of many modelers can be aggregated. New models can be integrated into existing systems without requiring the entire system to be redesigned. This enables the formulation of CRS as a discipline.

Third, the framework ensures traceability and auditability. Every part of a model can be traced back to a verse (Step One) and a tafsīr-grounded semantic statement (Step Two). Variables are drawn from a shared ontology (Step Three), and mathematical formulations are documented and inspectable (Step Four). Thematic labels and integrations (Step Five) can be reviewed and, if necessary, revised. This traceability is crucial for a discipline that deals with sacred texts. It allows scholars to ask, "Why did you model this verse this way?" and to receive an answer that is grounded, step-by-step, in the pipeline.

Fourth, the framework is technology-agnostic yet AI-ready. It does not assume a particular programming language, AI platform, or mathematical toolkit. What it does assume is that any system engaging Revelation computationally must respect the sequence of preserving text, grounding meaning, standardizing variables, formalizing structure, and integrating globally. As AI advances, models can become more sophisticated, simulations more realistic, and reasoning engines more powerful – but the need for this disciplined pipeline will not disappear. In fact, it becomes more important the more powerful our tools become.

Fifth, the framework is cross-scripture compatible. Although we are applying it here to the Quran, there is nothing in the structure of the five steps that limits them to Islam. Other religious traditions can build their own source layers, exegesis summaries, variable ontologies, formal models, and thematic maps. The result would be parallel computational atlases of different scriptures, allowing CRS to become a truly comparative discipline. The Quran serves as the first fully modeled case study, but the framework is intentionally designed to generalize.

Finally, the framework embodies a certain philosophy of modeling revelation: that one can be both reverent and rigorous; that modeling does not have to mean reduction; that structure can be made explicit without claiming total comprehension. It is a way of saying that divine guidance has patterns we can study, connections we can map, and systems we can analyze, all while acknowledging that the reality behind those patterns surpasses our equations.

CLOSING THOUGHTS

The Five-Step Modeling Framework may appear, at first glance, as a technical device – a sequence of operations for turning verses into mathematical objects. But it is more than that. It is a way of organizing our intellectual responsibilities when we bring powerful computational tools into contact with sacred texts.

Step One reminds us that we begin with the text itself, not with our abstractions. Step Two reminds us that we stand on the shoulders of centuries of exegesis, not outside it. Step Three teaches us to be explicit about what we are modeling and how we are carving up the space of concepts. Step Four forces us to be honest about the structure we claim to see, by writing it in a language that can be checked and executed. Step Five keeps us from getting lost in the details by placing every local model in a larger map.

In the chapters that follow, we will build on this framework in two directions. In the next chapter, we will unpack the mathematical techniques that appear in Step Four – functions, dynamics, thresholds, belief propagation, graph theory – and show how they illuminate the structure of scriptural guidance without overriding it. Then, we will turn to the Universal Quranic Variables themselves, the ontology that lies at the heart of Step Three, and explore how they make the Quran intelligible not only to scholars and readers, but to AI systems that can reason over its structure.

Chapter 6
Mathematical Techniques in Scripture Modeling

ABSTRACT

Chapter 6 explains the mathematical and formal techniques used in Computational Religious Studies (CRS) to express scriptural guidance as explicit, analyzable structures. Focusing on the fourth step of the Five-Step Modeling Framework, the chapter introduces functions, derivatives, integrals, thresholds, vectors, graphs, probability, information theory, and formal logic as tools for modeling belief, action, moral change, and social dynamics over time. These techniques are presented intuitively, showing how they naturally arise from Quranic patterns of condition, causality, accumulation, and transformation. Mathematics is framed not as a reduction of revelation but as a language of structural clarity, enabling simulation, comparison, and AI integration while preserving theological boundaries.

INTRODUCTION

The previous chapter established the Five-Step Modeling Framework as the core pipeline of Computational Religious Studies (CRS): we begin with the preserved text, move through tafsīr-grounded meaning, map that meaning into variables, formalize the structure mathematically, and then integrate each model into a global atlas. This chapter zooms in on that fourth step – mathematical and formal expression – and explains the major mathematical tools that are used throughout the Quranic modeling project.

The purpose here is not to turn the reader into a mathematician, nor to bury the subject under unnecessary rigor. The goal is clarity. We want the reader to understand what it means, in practice, to say that a verse is modeled as a "function," that "belief is treated as a trajectory," or that "guidance is represented by a vector field." We also want to show that none of this is arbitrary. The mathematics used in

DOI: 10.4018/979-8-3373-6940-2.ch006

CRS emerges naturally from the kinds of relationships the text itself describes – relationships between belief and action, between cause and consequence, between guidance and misguidance, between individual decisions and collective outcomes.

In other words, mathematics here is not a replacement for scripture, but a language for expressing its structure. The Quran is not random text. It exhibits patterns, constraints, feedback loops, and networks of meaning. The mathematical techniques introduced in this chapter exist to make those patterns explicit in a form that can be analyzed and, eventually, computed. Many of these techniques are visible in the detailed modeling work already completed in the Quranic Modeling Project, which uses functions, derivatives, integrals, vectors, entropy, and network structures to give formal shape to verse-level models.

Because this chapter is meant as an overview rather than a technical manual, we will move from simpler concepts to more complex ones. We begin with basic notions like variables and functions, then move through calculus (derivatives and integrals), thresholds and piecewise behavior, vectors and linear algebra, graph theory, probability and belief propagation, information-theoretic ideas, and formal logic. At each stage, the emphasis will be on intuition: what the tool means, why it is useful, and how it connects to scripture modeling – not on long proofs or dense notation.

MATHEMATICAL LANGUAGE AS A TOOL FOR STRUCTURE

To understand why mathematics is appropriate here, it helps to recall what modeling does. When we model a system, we are not claiming to be the system. We are selecting certain aspects of it – its variables and relationships – and representing them in a compressed, structured way. A map is not the territory, but a good map preserves enough structure to navigate meaningfully.

Scripture presents a universe in which certain things reliably lead to others: belief leads to specific actions; certain actions lead to forgiveness or punishment; persistent habits affect the heart; communities rise and fall based on justice or injustice; future outcomes are tied to patterns in the present. That is the language of structure and dynamics. Mathematics is simply the most precise language we know for expressing such structure.

Crucially, we are never modeling the divine essence. We are modeling the logic of guidance as it is presented to human beings. When a verse declares, for example, that those who meet certain conditions will obtain a particular outcome, that is structurally analogous to a conditional function: if these inputs are present, this output follows. When a passage speaks of hearts hardening over time or of societies decaying after a sequence of moral failures, that is structurally analogous to a dynamical system: a state variable changing as time progresses.

This chapter will therefore speak freely in terms of variables, functions, derivatives, and networks, but the boundary is always clear: we are formalizing the human-facing side of revelation – its conditional structures, its dynamics, its connections – not the hidden mechanics of the divine.

VARIABLES AND FUNCTIONS: THE BASIC BUILDING BLOCKS

The most basic mathematical objects used in CRS are variables and functions. A variable is a quantity that can take on different values. A function is a rule that links variables together, assigning outputs to inputs.

In the Quranic modeling project, variables are not arbitrary. They are standardized in the Universal Quranic Variable Ontology (UQVO), which defines re-usable categories such as belief, hypocrisy, charity, justice, oppression, gratitude, fear, patience, and many others. Each of these variables has a meaning anchored in tafsīr and a mathematical domain: some are continuous numbers in the interval ([0,1]), others represent counts or intensities, others are binary (present or absent). This shared ontology allows models from different verses to "speak the same language."

Once variables exist, the simplest form of a mathematical model is a function such as

$$O = f(I, A)$$

where I might represent an internal state (like belief), A might represent an action (like charity or injustice), and O might represent an outcome (like guidance, punishment, or social stability). The function f encodes how these inputs combine: it might be additive, multiplicative, thresholded, or something more complex.

Functions can be static or dynamic. A static function describes relationships at a single moment: given these states and actions, what outcome is associated with them? A dynamic function describes how a variable changes over time, responding to other variables and its own past values. We will return to this dynamic case when discussing derivatives and differential equations.

Scripts in the modeling process often use functions to represent the conditional logic of a verse – sometimes in very compact form. For example, an eligibility condition might be written as

$$O = 1\left[I \geq \tau_I \, and \, A \geq \tau_A\right]$$

where $1[\cdot]$ is an indicator function: it outputs 1 if the condition in brackets is true, and 0 otherwise. Here τ_I and τ_A are thresholds for belief and action that must both be met for the outcome to be active. This is a direct formalization of an "if and only if these conditions hold" statement.

One of the strengths of functions is that they allow us to articulate the type of dependence between variables: whether it is linear or nonlinear, whether the effect of a variable saturates at some point, whether small changes in input produce small or large changes in output. In a scriptural context, this matters because some relationships are gentle and continuous, while others are abrupt and qualitative. Mathematical functions give us the language to distinguish those cases.

TIME, CHANGE, AND THE DERIVATIVE

Scriptural content is not only about static states – it is deeply concerned with change over time. People grow in faith or fall away. Hearts soften or harden. Communities reform or collapse. The notion of *trajectory* is therefore central. To capture that, CRS makes use of the most fundamental tool from calculus: the derivative.

Informally, the derivative of a function with respect to time, written $\frac{dX}{dt}$, tells us how fast the quantity X is changing at a given instant. If $X(t)$ represents belief at time t, then $\frac{dX}{dt}$ describes whether belief is increasing (positive derivative), decreasing (negative derivative), or stable (derivative near zero).

In the modeling process, belief and other internal states are sometimes treated as continuous functions of time and governed by differential equations – equations that relate the derivative of a variable to its current value and to other influences. A simple example is

$$\frac{dI}{dt} = g\left(I, A, C\right)$$

where g encodes how belief changes based on current belief I, actions A, and context C. If repeated patterns of obedience and remembrance strengthen belief, then g will be positive when those actions are present. If persistent sin and heedlessness weaken belief, then g will be negative when those factors dominate.

In some models, this logic is extended to second-order differential equations, where not only the rate of change but the acceleration of change matters. For example, belief might respond to revelation and temptation in a way mathematically analogous to an oscillating system with damping, capturing the idea that some people respond to guidance with temporary enthusiasm that fades unless reinforced. These second-order equations can capture overshoot, instability, or convergence toward a stable level over time.

Derivatives also appear in more abstract forms, such as the gradient ∇, which measures the direction and rate of fastest increase of a multivariate function. When modeling concepts like "deviation of the heart" or "entropy of denial," gradients can express how quickly that quantity intensifies as one moves through a space of states or beliefs.

The ethical point of using derivatives is simple: many scriptural themes are about *movement*. People are not frozen at one moral coordinate; they are on paths. Derivatives give us language for those paths – how quickly someone is moving toward or away from guidance, how rapidly a society is sliding into corruption, or how responsive a heart is to a moment of repentance.

ACCUMULATION AND MEMORY: INTEGRALS AND SUMMATION

If derivatives are about instantaneous change, integrals are about accumulation. The integral of a function over time captures the total "area under the curve" – the cumulative effect of whatever that function represents.

In many models, integrals are used to represent the way actions or influences accumulate over a lifetime or over a historical period. For example, if $A(t)$ represents a rate of charitable giving or injustice at time t, then

$$\int_{t_0}^{T} A(t)\,dt$$

represents the total amount of that action between times t_0 and T. This is more faithful to scriptural themes that speak of "records" of deeds or of long-term patterns being weighed, rather than isolated snapshots. Again, the integral is not claiming to reproduce the metaphysical accounting of deeds; it is giving form to the idea that repeated actions accumulate into an enduring moral profile.

Integrals also appear in models of memory and decay. For instance, an equation might express a guidance potential $G(t)$ at time t as a weighted integral of past scriptural exposure:

$$G(t) = \int_0^t \kappa S(\tau)\, e^{-\lambda(t-\tau)}\, d\tau$$

where $S(\tau)$ is the intensity of scriptural input at past time τ, κ is a scaling factor, and $e^{-\lambda(t-\tau)}$ introduces an exponential "forgetting" effect: older exposures count less than recent ones because of human forgetfulness or resistance.

In discrete time, integrals are often replaced by summations. For example, if a model works in steps (events, days, or episodes), then the cumulative effect might be written as

$$M_n = \sum_{i=1}^{n} B_i$$

where each B_i is a blessing, warning, or trial at step i, and M_n is some cumulative memory or moral inertia after n such events. Summation makes it easy to express ideas like "many favors have been shown, yet gratitude has not kept pace" or "repeated injustices have slowly built up an oppressive system."

Another important use of integrals is in representing instantaneous events by means of the Dirac delta $\delta(t - t_0)$, which is zero everywhere except at a particular time t_0 but integrates to one. In the models, such an idealized impulse is used to represent events that are effectively instantaneous from the human perspective – sudden repentance, a decisive historical intervention, or the arrival of a final judgment moment that changes the state of a system.

Together, derivatives and integrals allow CRS to express both moment-by-moment changes and the cumulative effects of entire histories. They turn scriptural narratives of growth, decline, and remembrance into explicit trajectories that can be studied and compared.

THRESHOLDS, REGIMES, AND NONLINEAR BEHAVIOR

Many scriptural statements are not about gradual shifts but about qualitative changes that occur once a boundary is crossed. Hearts become "sealed" after persistent rejection; a community passes a point after which punishment becomes unavoidable; an act turns from permissible to forbidden once a certain combination of conditions is met. These are threshold phenomena, and they require mathematical tools beyond smooth, linear relationships.

In CRS, thresholds are modeled using piecewise functions and indicator functions. A simple threshold function might look like

$$S(x) = \begin{cases} 0, x < \tau, \\ 1, x \geq \tau, \end{cases}$$

where x might represent the intensity of a recurring sin or the cumulative degree of moral decay, and τ is a threshold beyond which a new regime starts – say, a sealed heart or a failure state. The indicator function $1[x \geq \tau]$ is a compact way to write the same idea.

Thresholds also allow us to model phase transitions: regimes in which the same input will have different effects depending on which side of the boundary one is on. For instance, a reminder or warning might have a strong positive effect on someone who is still receptive but almost no effect on someone whose receptivity has already dropped below a critical level. In formal terms, the derivative $\frac{dI}{dt}$ might switch from positive to near zero once I falls under a threshold.

Another class of nonlinear behavior arises through saturating functions like the logistic function

$$L(x) = \frac{1}{1 + e^{-k(x-x_0)}},$$

which grows quickly in the middle but levels off near 0 and 1. Such functions can represent processes that accelerate once they get going but cannot exceed certain bounds: spiritual growth that becomes easier after a habit is established, or corruption that spreads rapidly after a tipping point but then saturates as it envelops a community.

Thresholds and nonlinear functions thus capture the reality that spiritual and moral systems are not always smooth and incremental. There are tipping points, drastic turns, and qualitatively different states. Encapsulating those in explicit functions allows CRS to model why some warnings arrive "too late" or why small early differences in behavior lead to very different long-term paths.

Vectors and Linear Algebra: Multidimensional Moral Space

Scriptural ethics is rarely one-dimensional. Righteousness is not a single scalar; it is a composite of belief, worship, justice, compassion, patience, honesty, and more. Likewise, social well-being involves justice, solidarity, economic fairness, and absence of oppression. To represent such multidimensional states, CRS uses the language of vectors and basic linear algebra.

A vector is simply an ordered list of numbers, which we can think of as coordinates in a space. For example, a "righteousness vector" V_R might include components for belief in God, prayer, charity, patience, justice, and other variables:

$$V_R = [B_{God}, prayer, charity, justice, patience, ...]$$

The length (or norm) of this vector, often written $|V_R|$, can be used as an overall measure of ethical completeness, while individual components allow more refined analysis. In some models, verse-level descriptions of "true righteousness" are represented as such vectors, making it possible to compare different profiles and to study how certain components interact.

Vectors are also used for things like divine attributes, states of devotion, or social indicators. Once variables are bundled into vectors, we can use operations like dot products and angles to reason about alignment. If two vectors point in the same direction in moral space, their dot product is positive and the cosine of the angle between them is close to 1, indicating alignment. If they point in opposite directions, the dot product is negative and the angle is close to 180 degrees, indicating conflict. This idea appears explicitly in models that treat prophetic guidance and human rejection as opposing vectors, with their misalignment signaling opposition.

Linear algebra also provides tools like matrices, which can represent transformations from one vector to another. For example, a matrix might encode how a set of actions translates into changes across several internal states, or how group-level events update multiple components of social cohesion at once. While the Quranic modeling process uses matrices less heavily than vectors and scalar functions, the conceptual foundation is the same: linear transformations allow complex relationships to be represented compactly and applied repeatedly.

The advantage of a vector-based view is that it respects the richness of ethical and spiritual descriptions instead of reducing everything to a single number. At the same time, it provides structure: we can compare vectors, compute norms, analyze trends, and ask whether certain combinations of components are emphasized or neglected in particular contexts.

NETWORKS AND GRAPH THEORY: SCRIPTURE AS AN INTERCONNECTED SYSTEM

Another central mathematical tool in CRS is graph theory, the study of networks. A graph consists of nodes (also called vertices) and edges connecting them. In scripture modeling, nodes can represent many things: concepts, variables, agents, communities, or even verse-segments. Edges represent relationships: causal links, obligations, flows of trust, transmission of knowledge, or chains of inheritance.

In the Quranic modeling project, graphs are used in several ways. Conceptual graphs connect variables like belief, action, guidance, and punishment in directed structures, making it easy to see how they influence one another. Social graphs represent trust relationships, obligations, or debts between individuals, enabling analysis of risk distribution and fairness in economic dealings. Covenant graphs track the transfer of prophetic responsibility through lineages, while trust networks depict how obligations and risk are distributed in financial and social systems.

Once a system is represented as a graph, a rich set of analytic tools becomes available. We can study centrality (which nodes are most connected and thus most influential), paths (how an influence or responsibility travels from one node to another), clusters (subgroups that are closely linked), and bottlenecks (points where failure or corruption would have outsized effects). This helps make precise the intuitive sense that some concepts or duties are structurally central in scripture.

Graph theory also plays a role in modeling moral and spiritual allegiances. One simple example is the modeling of alternative "governance systems" as distinct attractor graphs: one centered around divine guidance and light, another around oppressive structures and darkness. Each graph has its own internal edges – flows of support, commands, and influence – and agents position themselves in one or the other. The graph formalism makes it clear that allegiance is not only about individual feelings but about embedding oneself in a particular network of relationships.

At a higher level, graph theory supports the global integration step of CRS. Each verse-level model can be seen as a local subgraph, and these are stitched together through shared variables and themes into a large, scripture-wide network. This is how CRS demonstrates that the Quran is not a random list of isolated statements but a densely interconnected epistemic system.

PROBABILITY, RISK, AND BELIEF PROPAGATION

Up to this point, we have mostly discussed deterministic relationships: given certain inputs, a model produces certain outputs. However, much of real human life is suffused with uncertainty. People are unsure about the unseen, about the future, and about the states of other people's hearts. Communities form expectations based on partial information. CRS therefore uses probability theory and related concepts to model uncertainty and risk in a disciplined way.

A probability is a number between 0 and 1 that represents the plausibility of an event or proposition, given current knowledge. In scripture modeling, probabilities can represent, for example, an agent's subjective confidence in a doctrinal claim, the likelihood that a deceitful person is lying, or the perceived chance that a risky but righteous action will be rewarded. Models may include functions such as $P(E|B)$, the probability of an event E given a belief state B.

A central tool here is Bayes' theorem, which describes how beliefs should be updated when new evidence arrives. In abstract form,

$$P\left(H\middle|E\right) = \frac{P(E|H)P(H)}{P(E)}$$

where H is a hypothesis (such as "this guidance is truly from God") and E is new evidence (such as a fulfilled promise, a witnessed sign, or a historical pattern).

The theorem provides a framework for thinking about how exposure to signs and experiences should, in principle, change a person's degree of belief.

In some models, probability appears more directly in decision theory or risk analysis. For instance, substance use or economic schemes may be conceptualized as having both benefits and harms, each with associated probabilities. An ethical decision rule can then be expressed in terms of expected value: if the expected harm outweighs the expected benefit – especially when measured in spiritual terms – then the system classifies the behavior as forbidden.

Beyond individual cognition, probability plays a crucial role in belief propagation over networks. Imagine a social graph where each node has some level of confidence in a belief, and edges represent trust or influence. As time passes, beliefs spread and change according to simple update rules: people adjust their beliefs toward those of the people they trust, weighted by their own prior convictions and the strength of the relationships. This picture is mathematically captured by models such as Markov chains or Bayesian networks, where probabilities flow through a graph.

In the context of scripture, this allows CRS to formalize phenomena like the spread of hypocrisy, the reinforcement of unbelief through collective denial, or the strengthening of faith in communities where trust networks are built around shared commitment. The mathematics here does not decide what is true theologically; it explores how human beliefs about truths can move through populations.

INFORMATION, SIGNALS, AND FILTERS

Another set of ideas that appears in scripture modeling comes from information theory and signal processing. In several models, revelation is treated as a signal – a structured stream of guidance – while human hearts and minds are treated as filters that may pass, distort, or block that signal. This symbolic representation proves fruitful for structuring the dynamics of guidance and misguidance.

A basic tool here is the notion of convolution, which describes how an input signal $S(t)$ is transformed by a filter with impulse response $h(t)$:

$$R(t) = \int_{-\infty}^{t} S(\tau)h(t-\tau)d\tau.$$

The function $R(t)$ is the response. In CRS, $h(t)$ can be thought of as encoding how a given person or community processes episodes of guidance: how quickly they respond, how long they remember, whether they are more sensitive to certain types of messages than others. A very "hard" heart might have a response function that heavily attenuates the signal, producing a weak or delayed response no matter how strong the input. A receptive heart might amplify certain aspects of the signal.

Information theory enters through quantities such as entropy, which measures uncertainty or disorder in a distribution. One common formula is

$$H(p) = -\sum_i p_i log(p_i),$$

where p_i are probabilities over different possibilities. In CRS, entropy can represent the degree of confusion or self-deception in a person's belief system. When people cling to wishful stories about their immunity to consequences, the "entropy of denial" can be modeled as high: their internal probability assignments are scattered across competing, comforting illusions rather than concentrated on what revelation actually says.

Additionally, ideas like minimum description length (MDL) and compressibility are used to articulate why a revealed text might resist human imitation: if no human-produced text can match its combination of brevity, richness, and structural coherence without being more complex in an underlying formal language, that difference can be expressed in MDL terms. Again, the point is not to reduce a sacred text to an encoding, but to show that certain claims about its uniqueness can be given a systematic formal interpretation.

Signal and information concepts therefore allow CRS to speak about revelation as a structured input, human cognitive and spiritual states as filters, and denial or acceptance as transformations of information. They make precise the ways in which messages can be distorted, forgotten, amplified, or rejected – ideas the Quran repeatedly emphasizes.

Logic and Formal Reasoning

Alongside numerical mathematics, CRS relies on formal logic to express the structure of commands, prohibitions, permissions, conditions, and consequences. Logical modeling is particularly useful because scriptural statements often have the form "If X, then Y," "X is forbidden," or "Y is required," regardless of whether numeric values are present.

At the most basic level, we work with propositional logic, where propositions (claims that can be true or false) are connected by operators like AND, OR, NOT, and IMPLIES. For example, the statement

- If a person believes and acts justly, then they are of the successful.

can be represented as

$$\forall x \, [Believer(x) \wedge Just(x) \rightarrow Successful(x)].$$

Here we have already moved into predicate logic, which allows us to quantify over individuals (using "for all" ∀ or "there exists" ∃ and to apply predicates like "Believer" or "Just" to them.

Logic becomes especially powerful when combined with deontic operators, which capture normative concepts:

- $O(\phi)$: it is obligatory that ϕ.
- $F(\phi)$: it is forbidden that ϕ.
- $P(\phi)$: it is permitted that ϕ.

A scriptural rule like "You must not consume stolen wealth" can therefore be written as $F(ConsumeStolenWealth)$. If there are exceptions or contextual qualifiers, they can be added as conditions: "If circumstance C holds, then $P(\phi)$; otherwise $F(\phi)$."

Logical formulations are used throughout CRS to represent legal structures, moral norms, and conditional promises or threats. They complement the numeric models by making explicit the hard boundaries and logical dependencies that are not easily expressed in continuous functions. They also allow for consistency checking: we can verify that no combination of rules forces contradictory obligations or permissions under the same circumstances.

When logic and mathematics are combined, we obtain hybrid statements such as: "If belief I is above threshold τ_I and action A remains above τ_A over the interval $[t_0, T]$, then it is guaranteed that outcome O will occur." Here the logical "if–then" wraps around inequalities and temporal conditions, capturing both normative and quantitative aspects of the scriptural statement.

In the context of AI, these logical structures can be implemented as rule sets that guide reasoning, ensuring that a machine working inside a CRS framework obeys the same constraints that the modeled scripture imposes on human moral logic.

HYBRID MODELS AND SIMULATION

In practice, the most informative models in CRS are hybrid: they combine several of the techniques discussed above into composite systems. A single verse or theme might be represented by a logical skeleton, a network of variables, differential equations describing their temporal evolution, and threshold functions capturing irreversible transitions.

For example, a model of long-term moral decline in a community might use graph theory to represent social ties, probability theory to represent the spread of certain attitudes, differential equations to express how individual beliefs and fears change over time, and thresholds to indicate when a community crosses a line that triggers collective consequences. Logical rules could then overlay this system, enforcing restrictions like "This kind of action is never permitted" or "This outcome only becomes relevant after a certain condition is met."

These hybrid systems are where simulation becomes possible. Once variables, equations, and rules are defined, one can run the model forward under different scenarios: what happens when reminders become more frequent, when unjust leaders gain more connections in the graph, when charity in times of hardship increases, or when knowledge is systematically concealed? Such simulations can illuminate the structural implications of the relationships encoded in scripture.

Simulation also helps with sensitivity analysis – seeing which variables a system is most sensitive to. If a small change in some parameter (say, the rate at which people forget blessings) dramatically shifts the long-term outcome, that suggests the corresponding spiritual quality (gratitude, in this case) is structurally central in the scriptural system. Mathematics thus becomes a magnifying glass for priorities that the text already points to.

THE QURAN AS AN ORGANIZED EPISTEMIC SYSTEM

By this point, the pattern should be clear. Functions express basic input–output relationships; derivatives and integrals express change and accumulation; thresholds express qualitative shifts; vectors handle multidimensional virtue and vice; graph theory structures networks of meaning and obligation; probability and information theory capture uncertainty, risk, and distortion; logic articulates rules, commands, and constraints. When all of these are anchored in a shared ontology of variables and in tafsīr-grounded semantics, the result is not chaos but coherence.

What emerges from this mathematical perspective is a picture of the Quran (and, by extension, other scriptures) as an organized epistemic system. It is a modeled coordinated set of claims about reality, human psychology, social dynamics, and ultimate consequences. The models reveal structured feedback loops – between belief and action, between gratitude and blessing, between injustice and downfall. They reveal stable configurations – states of heart and society that are repeatedly praised or condemned. They reveal thresholds and tipping points, where trajectories suddenly bend.

The mathematics itself does not create this structure; it reveals and codifies it. And because the models are explicit, they are both auditable and computable. A theologian can ask whether a particular function or threshold really reflects the best reading of the text. A computer scientist can implement the model in an AI system and test its behavior under different conditions. A philosopher can ask what kind of rational agent emerges from acting according to these modeled constraints.

This is precisely what it means for the Quran to be "suitable for advanced analysis" in the context of CRS. We are not treating scripture as just another dataset for a machine-learning algorithm to pattern-match. We are treating it as a source of structured guidance, and we are building models that respect that structure.

CLOSING REFLECTIONS

This chapter has walked through the main mathematical techniques used in Scripture Modeling: variables and functions as the foundational language; derivatives and integrals to capture change and memory; thresholds and nonlinearities to represent regime shifts; vectors and linear algebra to express multidimensional ethical states; graph theory to organize networks of concepts and agents; probability and belief propagation to deal with uncertainty and social influence; information-theoretic ideas to model signals, filters, and entropy; logic to encode rules and norms; and hybrid models that combine these elements into dynamic, analyzable systems.

Throughout, the emphasis has been on structural meaning rather than numerical obsession. The numbers and symbols are tools. They exist to make explicit what is otherwise implicit: that revelation speaks in patterns, trajectories, and networks; that it presents laws; and that those laws can be written in a language that both humans and machines can reason about, without ever confusing the model with the reality it points to.

In the next chapter, we turn to the other half of this modeling equation: the variables themselves. The Universal Quranic Variable Ontology (UQVO) is the semantic backbone that all of these mathematical techniques depend on. If this chapter has explained the grammar of the modeling language, the next will describe its vocabulary – the carefully defined concepts that bridge between scripture, human interpretation, and formal systems.

Chapter 7
The Universal Quranic Variables

ABSTRACT

Chapter 7 introduces the Universal Quranic Variable Ontology (UQVO) as the semantic backbone of Computational Religious Studies, providing a standardized, reusable vocabulary that allows Qur'anic concepts to be modeled consistently across verses, themes, and mathematical systems. The chapter explains why formal ontology is essential for large-scale scripture modeling and details how the Universal Quranic Variable Dictionary (UQVD) organizes divine attributes, human states, spiritual dynamics, group identities, temporal realities, and system operators into machine- and human-readable variables. By grounding each variable in tafsīr while defining its mathematical role, domain, and relationships, UQVO transforms recurring scriptural themes into interoperable components of a unified epistemic system. This ontology enables coherent modeling, AI integration, comparative analysis, and governance, positioning the Qur'an as a structured conceptual universe rather than a collection of isolated verses.

INTRODUCTION

In the previous two chapters we built the scaffolding of Computational Religious Studies (CRS). The Five-Step Modeling Framework established how a verse moves from preserved text to tafsīr, from tafsīr to variables, from variables to formal equations, and from equations to a global scripture-wide map. The chapter on mathematical techniques then showed what kind of tools we use in Step Four: functions, derivatives, integrals, thresholds, networks, and logical systems.

This chapter now turns to the element that sits in the middle of everything: the Universal Quranic Variable Ontology (UQVO). If the Five-Step framework is the pipeline, UQVO is the vocabulary that flows through it. If the mathematical techniques

DOI: 10.4018/979-8-3373-6940-2.ch007

provide the grammar of our formal language, UQVO provides the nouns and verbs. It is a standardized and evolving taxonomy of core Quranic concepts, expressed as variables that can be used consistently across verses, surahs, themes, and models. It is both human-readable and machine-readable, both scholarly and computational.

The Universal Quranic Variable Ontology is implemented, in its concrete form, as the Universal Quranic Variable Dictionary (UQVD), which lists each variable with a code, an Arabic term, a concise meaning, and a symbol for use in formulas. The UQVD is divided as follows: divine attributes labeled as A_X , human states as H_X, spiritual dynamics as D_X, group identities as G_X, temporal and eschatological concepts as T_X, and a small set of operators and system functions that tie them together.

In this chapter we will unpack that ontology: how it is structured, what principles guide it, how individual variables are defined, and how the ontology undergirds the entire CRS project. We will see why concept standardization is essential for any large-scale Quranic modeling effort, how the UQVO becomes a kind of semantic backbone for the Quran, and why it is central to making the revelation intelligible to AI systems without sacrificing theological integrity.

1. FROM WORDS TO VARIABLES: WHY AN ONTOLOGY IS NEEDED

To understand the role of UQVO, it helps to pause on a simple observation: the Quran is full of concepts that reappear in different contexts, with different shades of emphasis, but unmistakable continuity. Belief and disbelief, patience and arrogance, justice and oppression, guidance and misguidance, mercy and punishment, dunya (world) and ākhirah (afterlife) – these are not isolated terms. They are recurring nodes in a web of meaning.

Traditional tafsīr treats them as such. A scholar reading the entire Quran will see how "īmān" in one passage picks up themes introduced earlier, how "taqwā" deepens as a concept across surahs, how "hypocrisy" in the Medinan verses elaborates what was implicit before. The human mind can track these connections implicitly, drawing on memory, intuition, and years of study.

But a computer, or even a formal mathematical model, cannot rely on intuition. If every verse model uses its own labels – "faith" here, "belief level" there, "spiritual conviction" somewhere else – then the system as a whole becomes incoherent. We would have many local models, but no shared conceptual backbone. Concepts would fragment across the corpus, and any attempt at cross-verse or cross-surah analysis would collapse into ambiguity.

The Universal Quranic Variable Ontology exists to solve this problem. It standardizes concepts. Instead of treating every mention of belief as a separate, free-floating idea, we map them onto a single variable: H_I, "Īmān," a human state of faith or belief. Instead of letting each model pick its own representation for divine mercy, we assign a canonical variable A_R, representing "ra mah," the attribute of mercy or compassion. Instead of ad-hoc labels for "believers as a group," we define G_M as the group identity for "mu'minūn," those who meet certain beliefs and behavior criteria.

In other words, UQVO transforms recurring Quranic themes into a taxonomy of variables. Each variable is conceptually grounded in tafsīr but mathematically defined so that it can be plugged into the models described in earlier chapters. Once this ontology exists, each verse is no longer modeled in isolation. Rather, its model is built out of reusable conceptual pieces, each of which reappears across the text.

Ontology, in this context, means more than a list of terms. It means that the variables have types, relationships, and constraints. Belief is not treated as just another number; it is a human state variable with a particular domain, particular dynamics, and particular connections to other variables like taqwā (God-consciousness), hypocrisy, and guidance. Mercy is not an arbitrary label; it belongs to the divine attribute family and plays a specific role in functions that model forgiveness, providence, or delay of punishment.

Thus, UQVO is what allows CRS to move from a collection of verse-level analyses to an integrated epistemic system. Without it, the Five-Step Framework would generate many different models, but they would not be speaking the same language. With it, those models become interoperable and comparable, and the Quran's internal structure can be studied at a systemic level.

2. UQVO AND UQVD: ONTOLOGY AND DICTIONARY

It is helpful to distinguish between the ontology and its dictionary. The Universal Quranic Variable Ontology (UQVO) is the abstract schema: the conceptual categories, relationships, and structural rules that govern how variables are defined and how they relate. The Universal Quranic Variable Dictionary (UQVD) is one concrete instantiation of that ontology: a catalog of specific variables with identifiers, Arabic terms, concise meanings, and notational forms.

For example, the ontology says that there is a class of variables called "Divine Attributes," abbreviated A_X, representing aspects of the divine nature such as mercy, justice, power, knowledge, oneness, and self-sufficiency. It also says there is a class of "Human States," H_X, for inner conditions such as belief, disbelief, taqwā, hypocrisy, submission, doubt, and ego. Similarly, it defines "Spiritual Dynamics" D_X for processes like guidance, temptation, repentance, hardening of the heart, light, and barakah; "Group Identities" G_X for believers, disbelievers, hypocrites, people of the book, oppressors, and the righteous; and "Temporal / Eschatological" variables

T_X for worldly time, afterlife, day of judgment, lifespan, and eternity. The ontology then also includes a category of "Operators and System Functions," collecting meta-level constructs such as mapping functions, gradients, integrals of guidance, exponential decay under temptation, Dirac deltas at resurrection, and indicator functions for group membership.

The dictionary, in turn, lists specific instances within each category: A_R for ra mah (mercy), A_J for ʿadl (justice), A_P for qudrah (power), A_K for ʿilm (knowledge), A_U for a ad or wā id (oneness), A_S for amad (self-sufficiency), A_G for hidayah (guidance-giving capacity); H_I for īmān (faith), H_K for kufr (disbelief), H_T for taqwā, H_N for nifāq (hypocrisy), H_S for islām (submission), H_D for shak or rayb (doubt), H_E for nafs (ego/self); and so on.

By separating ontology and dictionary, CRS ensures both flexibility and stability. The ontology can be extended with new families or relationships as deeper modeling suggests, while the dictionary can be versioned and refined, signaling that this is an evolving artifact with controlled updates.

In practical terms, modelers use UQVD as their reference. When they encounter a concept in a verse during Step Three of the Five-Step Framework, they look for the corresponding variable code in the dictionary. If it exists, they use that; if not, they propose an addition to the ontology, which is then vetted against the overall structure. This disciplined process prevents concept proliferation and keeps the modeling ecosystem coherent.

3. DESIGN PRINCIPLES OF THE UNIVERSAL QURANIC VARIABLE ONTOLOGY

The UQVO is guided by a set of design principles that preserve both theological fidelity and computational usability. It is worth walking through these principles, because they explain why the ontology looks the way it does and what constraints it places on modeling.

The first principle is textual and tafsīr anchoring. No variable exists in abstraction. Every UQVO variable must be traceable to Quranic usage and supported by recognized tafsīr. This means that "īmān" becomes a variable not simply because it is a common word, but because the Quran repeatedly uses it as a central organizing concept, and the scholarly tradition has elaborated its meaning and boundaries. The same is true for "kufr," "taqwā," "nifāq," "ra mah," "ʿadl," and so on.

The second principle is semantic stability with internal richness. Each variable must represent a stable concept, but it does not attempt to exhaust every nuance. For instance, H_I (īmān) is defined as a state of internal affirmation and trust in divine truth, manifested in conviction and action. Within that variable, the models can represent intensity, fluctuation, growth, or decay, but the underlying referent remains anchored. This prevents the ontology from fragmenting each nuance into a separate variable, which would make cross-surah analysis impossible.

The third principle is orthogonality and non-redundancy. Variables are designed to be as non-overlapping as reasonably possible. Taqwā is not defined in a way that makes it indistinguishable from īmān; nifāq is not simply "low belief." Each variable captures a distinct dimension, even if those dimensions are deeply related. This orthogonality is what makes vector-based modeling of ethical states meaningful: when you bundle H_I, H_T, and H_N into a vector, each component actually adds new information.

The fourth principle is composability. Variables are defined so that they can be combined into higher-order structures: functions, vectors, graphs, differential systems. The ontology is therefore not a theological encyclopedia but a modeling toolkit. It focuses on concepts that naturally serve as inputs, outputs, or state dimensions in the kinds of mathematical models described in the previous chapter.

The fifth principle is machine readability without losing human intelligibility. Variable codes like H_I, A_R, or D_G are compact enough to appear in equations and to be parsed by software, but they are also interpretable by scholars. The dictionary preserves the Arabic term, an English gloss, and a symbol so that a line of formalism can be read both by a parser and by a human who recognizes the underlying concept.

Finally, the ontology is built with versioned extensibility. As CRS expands, new themes may demand new variables or refined distinctions. Instead of treating the ontology as frozen, it is treated as a living structure with clear versioning. New versions must keep backward compatibility where possible and document any changes that affect previous models. This makes UQVO sustainable as the core of a long-term research program.

4. ANATOMY OF A QURANIC VARIABLE

To appreciate the depth of UQVO, it is useful to zoom in on what a single variable actually contains conceptually. Earlier, we defined a formal schema for variable entries – identity, semantics, mathematical domain, logical role, dynamics, relationships, and textual anchoring. Here, we will walk through those ideas in prose, using one variable as an example.

Consider H_I, representing "īmān," the state of faith or belief. In the dictionary, it is listed under Human States, with the Arabic term "īmān," the meaning "Faith / Belief," and the symbol H_I. In the ontology, this variable has a canonical ID, which is used consistently in every model; a display name, such as "Belief (Īmān)"; and a connection to its classical roots in Arabic morphology and theology.

Semantically, the ontology defines H_I as an internal epistemic and volitional state of affirming divine truth with conviction, reflected in both inner assent and outward compliance. This definition is drawn from mainstream theological and tafsīr sources. It is narrower than the English word "belief," which could mean many things, and more specific than mere knowledge: it includes trust and willingness to act accordingly.

Mathematically, H_I is assigned a domain, often taken as the continuous interval $[0, 1]$. That domain represents the intensity or completeness of īmān, from complete absence (0) to an idealized upper bound (1). This does not mean that faith is literally a number in reality; it means that for purposes of modeling trajectories and thresholds, we represent it as such. Within that domain, we can then define derivatives $\frac{dH_I}{dt}$ to represent growth or decline over time, integrals to represent cumulative faith exposure, and thresholds τ_I to represent the difference between a merely nominal belief and an effective, salvation-bearing belief.

From a logical perspective, H_I plays certain roles. It appears as a condition variable in many Qur'anic statements: "those who believe and do righteous deeds..." can be formalized as requiring H_I above some threshold, combined with an action variable. It also appears as a trigger for certain promises and as an inhibitor for certain warnings. In the ontology, these roles are specified so that automated reasoning systems know how to treat H_I within rule structures.

Dynamically, H_I is treated as a variable that can change. The ontology flags it as "dynamic" rather than static, and models often attach differential equations to it. For example, in the dynamic belief model, the belief of different archetypes is governed by equations of the form

$$\frac{d^2 H_I}{dt^2} + \alpha \frac{dH_I}{dt} + \beta H_I = AG\left(t\right)$$

where $AG(t)$ is the guidance input and (α, β) are parameters capturing resistance or decay. This tells us that H_I is the subject of forces, not a fixed label.

Relationally, H_I is linked to other variables. It has positive affinity with H_T (taqwā), negative correlation with H_K (kufr) and H_N (nifāq), and functional relationships with D_G (guidance vector) and A_G (divine guidance capability). In the ontology, these are recorded as relationships so that the global knowledge graph can recognize, for example, that a verse invoking "hypocrites" involves both H_N and a derivative of H_I that oscillates rather than converges.

Finally, a variable like H_I is anchored in specific verses. The ontology links it to all the modeling units where it appears, creating a two-way mapping: from verse to variable and from variable to verses. This is essential for audits and for thematic analysis.

In short, a UQVO variable is more than a label. It is a small semantic and mathematical universe: a concept with a defined meaning, a numerical representation, a logical role, dynamic behavior, and explicit relationships, all traceable back to the Quran.

5. THE MAJOR ONTOLOGICAL FAMILIES

The UQVO organizes its variables into major families. Although the ontology may grow beyond these, the initial categories already capture much of the Quran's structure and have proven effective in the modeling work that has been done so far.

5.1 Divine Attributes A_X

The first family covers Divine Attributes, labeled A_X. These variables represent qualities of God that are frequently emphasized in the Quran and that play structural roles in guidance, law, mercy, and punishment. The dictionary lists A_R for ra mah (mercy/compassion), A_J for ʿadl (justice), for qudrah (power/might), for ʿilm (knowledge/omniscience), for a ad or wā id (oneness/uniqueness), for amad (self-sufficiency/refuge), and for hidayah (guidance-giving capability).

In formal models, these attributes often appear as parameters or fields rather than as changing states. For instance, in the modeling of Sūrah al-Fāti a, the opening verses are represented as a composite divine function

and the vector of divine attributes

is used inside a "praise function" that explains why praise arises as a response to recognition of these attributes.

Theologically, of course, these attributes are eternal and not subject to time. In CRS, they are therefore treated as exogenous constants in most models: they shape human trajectories without being shaped by them. From a modeling perspective, one can think of them as "global parameters" that define the structure of the universe – much like physical constants in cosmology define the behavior of a physical system.

The inclusion of these attributes in the ontology ensures that references to God's mercy, justice, power, knowledge, and oneness are not treated as free-floating adjectives but as structured components that interact with human variables. For example, and together influence forgiveness models: mercy tends to reduce punitive outcomes for repentant agents, while justice ensures that unrepented wrongdoing is not treated as morally neutral.

5.2 Human States

The second family consists of Human States, labeled . These are internal, often invisible conditions of the human soul: belief, disbelief, taqwā, hypocrisy, submission, doubt, and the ego. The dictionary lists for īmān, for kufr, for taqwā, for nifāq, for islām, for doubt (shakk/rayb), and for nafs.

These variables are central because many Quranic structures can be expressed as functions of human states and actions. In the dynamic belief models of Sūrah al-Baqarah, for instance, , and are treated as time-dependent functions representing the belief trajectories of believers, disbelievers, and hypocrites. Believers have a second-order differential equation driven by divine guidance; disbelievers are modeled as sealed – in effect, (= 0); hypocrites exhibit oscillatory behavior, never settling into stable convergence.

By defining these states as standardized variables, the ontology allows models from different parts of the Quran to refer to the same underlying dimensions. A passage about the "muttaqūn" in one surah and a passage about "those with taqwā" in another are both mapped to and related variables. A verse about doubt in revelation and a verse about doubt in resurrection both invoke , which can then be tracked across time and context. The same goes for the nafs, which interacts with temptation and hardening of the heart () in explicit operator expressions.

Human state variables are typically dynamic, subject to influence from revelation, temptation, social environment, and personal choice. They are therefore natural arguments in the functions described in Chapter 5, and the ontology defines them so that those functions are well-typed and interpretable.

5.3 Spiritual Dynamics ()

While human states describe "what is" inside a person at a given moment, the Spiritual Dynamics family, , describes processes and forces that act on those states over time. The dictionary includes (the hidayah vector, a directional force of guidance), (waswasa, temptation or whispering), (tawbah, repentance), (whispering from Satan), (qaswa, hardening of the heart), (nūr, light of understanding), and (barakah, a blessing or grace field).

In the operator section of the UQVD, several of these dynamics appear explicitly: the integral of over time represents accumulated guidance; the exponential represents decay of beneficial states under sustained temptation; the gradient expresses rates of change in human states; and the Dirac delta models instantaneous events such as the Day of Judgment.

These dynamics variables are crucial because they turn the ontology from a static taxonomy into a living system. Guidance is not just a status; it is an influence vector that can bend trajectories. Temptation is not just an idea; it is a force that can erode states of obedience. Repentance is not just a click of a switch; it is modeled as an impulsive correction in the control system of moral integrity.

By encoding these dynamics as variables, UQVO allows the mathematical techniques of Chapter 5 – differential equations, integrals, vector fields – to be applied in a spiritually meaningful way. We can write expressions likes

,

which say, in structured form, that belief changes under the combined action of guidance and temptation.

5.4 Group Identities

Scriptural discourse often refers not only to individuals but to groups: believers, disbelievers, hypocrites, people of the book, oppressors, the righteous, and others. The UQVO captures these as Group Identity variables, . The dictionary lists for mu'minūn (believers), for kāfirūn (disbelievers), for munāfiqūn (hypocrites), for ahl al-kitāb (people of the book), for ālimūn (wrongdoers/oppressors), and for āli ūn (righteous, sincere people).

Unlike human states, which are scalar or vector values, group identities are best thought of as sets or indicator functions. The operator section defines an indicator that returns 1 if an agent belongs to group and 0 otherwise. Membership in a group is typically defined in terms of thresholds or configurations of human state variables. For instance, being in might require and above certain thresholds, while remaining in might require oscillatory patterns in combined with outward conformity in actions.

These group variables allow models to capture the collective dimension of Quranic discourse. When a verse speaks of "the hypocrites" or "those who wronged themselves," the model can interpret that as referring to the set of individuals whose state vectors satisfy particular conditions. That, in turn, makes it possible to study how entire groups evolve over time or respond to events – a necessary step for modeling communal guidance, punishment, or reform.

5.5 Temporal and Eschatological Variables

Time is a central axis in the Quran: dunya versus ākhirah, this life versus the next, history versus eternity, the specific moment of the Day of Judgment versus the ongoing flow of worldly time. The UQVO captures these axes in the Temporal / Eschatological family, . The dictionary lists for dunya (worldly time), for ākhirah (afterlife), for yawm al-qiyāmah (Day of Judgment), for ayāt (lifespan), and for khālidūn (eternity/unending duration).

These variables play two roles. Conceptually, they allow models to mark which domain an outcome belongs to: a reward in dunya, a punishment deferred to ākhirah, a status that persists into eternity. Formally, they are often used to define time scales and events. For instance, the Dirac delta defines a singular time at which certain state transitions occur – a formal stand-in for the idea that at the Day of Judgment, the system undergoes an irreversible global transition.

Moreover, lifespan can be attached to individual agents, defining the interval over which their state variables evolve, while is used to mark outcomes that, in the models, are treated as terminal fixed states: once a trajectory enters an eternal reward or punishment state, the system no longer evolves. By making these temporal aspects explicit, the ontology supports models that take seriously the Quran's insistence on both short-term and long-term consequences.

5.6 Operators and System Functions

Finally, the ontology includes a set of Operators and System Functions that act on variables: $\Phi(H, A, t)$ as a mapping function from human state and divine attributes over time; as a gradient or rate of change in human state; as accumulated guidance influence; as exponential decay under sustained temptation; as a Dirac delta at the Day of Judgment; and as a group membership indicator.

These are not variables themselves but part of the ontology because they define the standard operations that CRS uses when building models. They form a kind of "API" between the variable world and the mathematical world: if a model wants to compute total guidance received, it is expected to use the integral of ; if it wants to represent temptation over time, it may use the exponential form; if it wants to switch behavior at the eschatological threshold, it uses .

Including these operators in the ontology ensures that different modelers don't reinvent incompatible ways of representing the same basic scriptural structures. It also means that AI tools working with these models can recognize patterns across verses: whenever they see , they know that some form of cumulative guidance is being represented, regardless of the specific context.

6. FROM LEXICAL ROOTS TO ONTOLOGICAL VARIABLES

One of the subtle strengths of UQVO is the way it separates lexical information from ontological structure. The Quran uses many related words with shared roots and overlapping meanings. For example, the root ʾ-m-n underlies words related to faith, security, and trust; k-f-r underlies disbelief and covering; q-s- and ʿ-d-l relate to justice. A naive approach might treat each distinct word as a separate variable. That would be linguistically precise but structurally useless: the system would explode into hundreds of variables with tiny differences, and the common architecture of guidance would be obscured.

UQVO takes a different route. It asks: what are the core system-level concepts that the Quran repeatedly uses to describe the moral and spiritual order? Once those are identified, it maps multiple lexical expressions into the same variable where appropriate. Thus, various conjugations and nominal forms of īmān map to ; various forms of kufr map to ; different phrases related to taqwā map to . The Arabic term in the dictionary and the symbol remind us of the underlying language, but the variable itself stands at a slightly higher level of abstraction, representing the concept across its linguistic realizations.

This move from lexeme to concept is crucial for machine readability. An AI system processing Arabic text can perform morphological analysis and then map root patterns to UQVO variables, thereby lifting raw language data into structured semantic form. From there, models can operate at the level of faith trajectories, temptation forces, group identities, and so on, rather than at the level of surface words.

At the same time, the ontology is careful not to flatten distinct concepts into one variable just because they share a root. For instance, "islām" (submission) and "salām" (peace) share letters but are conceptually distinct; so are "wa id" and "a ad" in some theological contexts. The ontology caps the abstraction at the level where structural roles stabilize, not where all lexical differences disappear.

In this way, UQVO becomes a bridge between the philological richness of the Quranic language and the structural clarity needed for computational models.

7. TAXONOMY AND GRAPH: VERTICAL AND HORIZONTAL STRUCTURE

The word "ontology" sometimes gives the impression of a static classification tree. UQVO is indeed a taxonomy – Divine Attributes, Human States, Spiritual Dynamics, Group Identities, Temporal/Eschatological – but it is also a graph, a network of relationships among variables. Both dimensions are important.

Vertically, the taxonomy helps us organize variables by type. We know that is a divine attribute, that is a human state, that is a dynamic process, that is a group, and that is an eschatological event. This vertical classification informs modeling: divine attributes are treated as exogenous constants; human states are modifiable internal variables; spiritual dynamics are forces; group variables define sets; temporal variables define timelines.

Horizontally, the ontology records the relationships between variables. For example, (guidance-giving capability) is linked to (guidance vector) as its operational manifestation; , in turn, acts on and . (hypocrisy) is defined partly as a misalignment between and outward actions and is associated with oscillatory dynamics in belief models. is defined as the set of agents whose and meet certain conditions, while is tied to patterns in . interacts with all these variables through the Dirac delta operator that freezes trajectories and outputs final outcomes.

When these relationships are visualized, UQVO becomes a conceptual graph of the Quran: nodes representing divine attributes, human states, dynamics, groups, and times; edges representing influence, dependency, or definitional links. That graph is then used in Step Five of the modeling framework to integrate verse-level models into a global atlas.

Because the ontology is shared, the graph has consistent semantics. A link between and in a model of Sūrah al-Fāti a means the same thing as a link between and in a model of Sūrah al-Baqarah. This cross-verse consistency is exactly what the ontology is designed to enable.

8. UQVO IN THE FIVE-STEP MODELING FRAMEWORK

Having examined UQVO in isolation, it is important to situate it inside the broader CRS methodology. The ontology is not an abstract catalog that sits on a shelf; it is actively used in the Five-Step Framework.

In Step One, modelers extract the source text of a verse in Arabic and translation. In Step Two, they produce a tafsīr-grounded semantic summary, identifying the key concepts, relationships, and conditions. Step Three is where UQVO enters explicitly. The modeler takes the semantic summary and asks: which variables from the ontology correspond to these concepts? Belief becomes ; mercy becomes ; guidance becomes and ; hypocrites become and ; dunya and ākhirah map to and ; temptations become ; and so on.

If the semantic content includes something not yet covered by UQVO, the modeler proposes a new variable, which then undergoes review. But most of the time, the dictionary already contains the necessary entries, precisely because typical Quranic themes were the starting point for UQVO's construction.

Once variables are mapped, Step Three also involves building conceptual models: specifying which variables are inputs, which are outputs, which are conditions, and how they are linked. Because variables are standardized, these local models automatically align with models from other verses that use the same ontology.

Step Four then turns these conceptual structures into equations, logic rules, graphs, or dynamic systems. Here, the variable codes from UQVO appear directly in the mathematics. A belief trajectory equation uses ; a guidance integral uses ; a group indicator uses ; a Day of Judgment event uses and its Dirac delta.

Finally, in Step Five, the modeled verse is thematically labeled and integrated into the global map. This integration uses UQVO again: themes are often defined in terms of variable clusters – "faith dynamics" involves ; "economic justice" involves , riba (usury), , , and so forth. Because the ontology is shared, the thematic map can span the entire Quran coherently.

In short, UQVO is the bridge between textual meaning and formal structure. It is what makes Step Three possible and Steps Four and Five consistent.

9. MACHINE READABILITY AND AI INTEGRATION

One of the explicit goals of CRS is to enable AI systems to engage with scripture in a way that is more than surface-level text prediction. For that to happen, AI needs access to structured representations of scriptural content – something like a world model. UQVO is the entry point for that world model.

In a computational setting, each UQVO variable can be represented as a symbolic identifier, a node in a knowledge graph, or a typed field in a structured data object. Its domain can be enforced by type systems or constraints (for example, $\in [0,1]$); its relationships can be encoded as edges or rules; its dynamics can be implemented as update functions. The UQVD's inclusion of symbols and concise meanings makes it straightforward to generate both human and machine representations from the same source.

For AI models, this means several things. First, natural language processing pipelines can map Quranic text to UQVO variables using morphological analysis and semantic parsing. When a verse mentions "those who believe," the system can recognize that as involving and ; when it mentions "day of judgment," it maps to . Second, reasoning engines can operate on these variables using the equations and rules developed in Chapter 5. For example, a system could answer questions like "What happens if a hypocrite hears more reminders?" by simulating the dynamic equations for and related variables.

Third, hybrid neuro-symbolic architectures can combine variable-level reasoning with vector embeddings learned from text. UQVO variables can be associated with embedding vectors that capture their usage contexts, while their formal relationships constrain how those embeddings may be combined. This allows AI to benefit from both statistical patterns and explicit semantic structures.

Most importantly, UQVO supplies a consistent reference layer that can be audited by scholars. When an AI system claims that a certain state of belief leads to a certain outcome, we can trace that claim back through UQVO to the verses and tafsīr that informed the model, and we can inspect whether the equations are faithful. Without such an ontology, AI reasoning over scripture would be opaque and unaccountable.

10. EVOLUTION, GOVERNANCE, AND VERSIONING

Given its centrality, UQVO cannot be treated lightly. It defines the conceptual primitives of the CRS universe. For that reason, the ontology is managed as a versioned and governed artifact. The versioning tag on the Universal Quranic Variable Dictionary is a signal of this.

As modeling work expands, new concepts may surface that deserve variable status: perhaps more specific social roles, finer distinctions between forms of hypocrisy, or additional temporal markers. When that happens, the ontology is not simply edited on the fly. Instead, proposed additions or modifications are evaluated against several criteria: whether the concept is genuinely recurrent in the Quran; whether it is well defined in tafsīr; whether it can be modeled in a way that adds structural clarity rather than redundancy; and whether its introduction maintains or improves the coherence of existing models.

Once accepted, new variables are assigned codes, placed into an ontological family, given domains and roles, and integrated into the dictionary. Older variables are rarely removed, but they may be refined or deprecated with clear change logs. Models built against earlier versions of UQVO can then be migrated when necessary, or kept as legacy layers for comparison.

This governance process also opens UQVO to broader scholarly participation. Although the initial ontology arises from one research team, its evolution can include contributions from Quranic scholars, mathematicians, computer scientists, and philosophers, provided they follow the foundational principles described earlier. In this sense, UQVO is not just a technical artifact but an emerging scholarly standard for representing Quranic concepts in computational form.

11. UQVO AND COMPARATIVE SCRIPTURE

While the Universal Quranic Variable Ontology is tailored to the Quran, the idea of a scripture-aligned variable ontology is not limited to Islam. Other religious traditions can, in principle, build their own ontologies – structured dictionaries of core concepts grounded in their own texts and exegetical traditions.

What UQVO offers, in this context, is a template. It shows how to move from text to variables, how to group concepts into families, how to define domains and roles, and how to build models that respect theological boundaries. It also provides a basis for comparative work.

For instance, UQVO's (īmān) could be aligned with a variable representing faith or trust in a Biblical ontology; (ra mah) could be aligned with a mercy variable; (Day of Judgment) with eschatological judgment concepts. The alignment need not be perfect, and differences are as important as similarities. But having explicit variables in each system makes comparison possible at a level deeper than word-to-word matching.

This opens the door to a truly comparative Computational Religious Studies: multiple scriptures modeled with their own ontologies, then linked by mappings that respect both their uniqueness and their shared concerns. UQVO thus serves not only the Quranic modeling project, but also the broader vision of CRS as a discipline that can analyze religious thought across traditions with both rigor and respect.

12. CONCLUSION: UQVO AS THE SEMANTIC BACKBONE OF CRS

The Universal Quranic Variable Ontology is one of the central achievements – and necessities – of the Quranic modeling project. It takes the recurring themes of revelation and turns them into a structured, standardized, and machine-readable vocabulary. It enables concept standardization across verses and surahs. It provides the taxonomy that organizes divine attributes, human states, spiritual dynamics, group identities, and temporal realities. It supplies the variables that the mathematical models of Chapter 5 operate on. It furnishes the nodes for the global graphs of Chapter 4's Step Five.

Without UQVO, CRS would remain a set of clever but disconnected models. With it, the Quran becomes visible as an organized epistemic system: a network of beliefs, actions, states, and outcomes that can be studied systematically while remaining anchored in its sacred text. The ontology allows both scholars and AI systems to speak a common language about that system and to interrogate its structure under the watchful guidance of tafsīr.

As the project grows, UQVO will continue to evolve. New variables will be added, relationships refined, themes expanded. But the core idea will remain: that revelation can be approached not only as a source of spiritual inspiration but also as a structured universe of concepts, amenable to careful modeling. In that sense, the Universal Quranic Variables are not just technical devices; they are the conceptual skeleton through which CRS seeks to understand, in structured form, how the Quran speaks about God, humanity, history, and destiny.

Section 3
Structural Modeling of the Quran

Chapter 8
Modeling Surah al-Fatiha: The Opening Function

ABSTRACT

Chapter 8 marks the transition from theory to practice in Computational Religious Studies (CRS) by presenting full mathematical models of Sūrat al-Fāti ah and the first thirty-nine verses of Sūrat al-Baqarah. Using the Five-Step Modeling Framework, the chapter demonstrates how familiar Qur'anic passages can be expressed as structured systems of variables, vectors, differential equations, networks, and logical constraints while preserving their theological depth. It models divine attributes, guidance as a directional field, belief as a dynamic trajectory, hypocrisy as oscillatory instability, and repentance as a corrective impulse. By grounding each formal choice in tafsīr and explaining its mathematical justification, the chapter reveals the Qur'an as a coherent, multi-layered epistemic system and provides a reusable template for modeling revelation at scale.

INTRODUCTION

MODELING SURAH 1 and FIRST 39 VERSES OF SURAH 2

In the earlier chapters you were introduced to Computational Religious Studies (CRS) as a field: its philosophical foundations, its relationship to classical tafsīr and theology, and its ambition to treat revelation as something that can be represented, analyzed, and even simulated using the tools of modern formal reasoning. Now we cross a threshold. In this chapter, we move from theory *about* CRS to the practice *of* CRS. We invite you into the mathematical workshop and show you actual models drawn directly from the Qur'an. You will see, step by step, how verses you already know – Al-Fāti ah and the opening forty verses of Sūrat al-Baqarah – can be expressed

DOI: 10.4018/979-8-3373-6940-2.ch008

as structured systems of variables, relationships, and equations without losing their theological depth or spiritual power.

Our aim here is not to impress you with symbols for their own sake, but to reveal the sheer richness of mathematical structures latent in the Qur'anic discourse. Just as economists, sociologists, and psychologists routinely turn to calculus, algebra, statistics, and dynamical systems to make sense of human behavior, CRS treats the Qur'an as a body of knowledge whose internal logic can be articulated using many branches of mathematics. When a social scientist speaks of "utility functions," "equilibria," or "feedback loops," they are not denying the humanity of their subjects; they are adding an additional layer of clarity. Similarly, when we model guidance as a vector, hypocrisy as an oscillatory signal, or repentance as an impulse in a moral control system, we are not reducing the Qur'an to math – we are expanding our ability to see its structure.

In the CRS corpus underpinning this book, we have drawn on a broad range of mathematical methods. We have used set theory to describe collections of states – such as sets of divine attributes, moral categories, or deviation patterns. We have employed algebra to express functional relationships between variables, for instance how devotion conditions divine aid. We have used linear algebra when the Qur'an speaks in multi-dimensional terms: when it layers attributes of God, components of faith, or interacting forces within a single conceptual object. We have turned to calculus and differential equations whenever the text clearly points to dynamics – growth, decay, oscillation, convergence, or divergence over time. We have even drawn on signal processing to represent recitation and remembrance as inputs flowing through cognitive "filters" in the human heart. Each of these choices emerges from the nature of what the verse is doing, not from any preconceived desire to "force" a particular type of math onto the text.

This chapter will focus on two foundational windows into the Qur'an: the entirety of Sūrat al-Fāti ah and the first thirty-nine verses of Sūrat al-Baqarah. These passages are not only central to Muslim spiritual life; they also form an ideal testbed for CRS because they already contain, in compact form, many of the major themes of the Qur'an – divine ontology, human dependence, guidance and misguidance, archetypes of belief, cosmological signs, and the first narrative of human moral responsibility in the story of Adam. By modeling these units in detail, we are in effect building a prototype "semantic engine" for the rest of the scripture.

Crucially, the mathematical structures you will see are not applied haphazardly. We do not decide in advance, "This verse will be calculus, that verse will be linear algebra," and then retrofit the meanings to match. The process runs in the opposite direction: we begin with a close reading of the verses, informed by tafsīr, and then ask a disciplined set of questions. What exactly is being related to what? Is the verse describing a *state* or a *trajectory*? Is it classifying types or describing flows and changes? Are we dealing with a single dimension or several interacting dimensions at once? Is the emphasis on accumulation over time, on equilibrium, on oscillation, on filtering, or on comparison between alternatives? The answers to these questions guide us toward the branch of mathematics best suited to express what the verse is already doing conceptually.

GENERAL METHODOLOGY

To keep this discipline visible and teachable, each modeled segment in this chapter will be presented through a common five-part lens. For every Surah 1 unit, and for each cluster of verses in Surah 2 up to 2:40, we will explain:

Variables – What are the core quantities or states in play? Here we name them explicitly and connect them to a shared dictionary of Qur'anic variables. You will see, for example, divine attributes like mercy and power, human states like faith, taqwā, doubt, disbelief, and hypocrisy, as well as structural objects such as guidance vectors, moral potentials, and epistemic filters.

Nature of the Relationship – What is the verse asserting about how those variables interact? Is it a direct dependency, a conditional promise, a comparative ranking, a dynamic push-and-pull, or a classification into types? At this stage we stay conceptual, clarifying the logical structure of the verse before writing a single symbol.

Why This Type of Math – Only once the conceptual relationship is clear do we justify the mathematical domain. If the verse describes change over time in response to inputs, we reach for calculus and differential equations. If it layers multiple components into a single conceptual whole, we treat them as vectors and use linear algebra. If it speaks of sets of people, paths, or states, we lean on set theory and logical predicates. If it's about signals, filters, and repeated exposure, we employ ideas from signal processing. This is where you will see the methodological reasoning that connects Qur'anic logic to mathematical form.

Why This Specific Model Form – Within a chosen mathematical domain, there are many possible models. Why a second-order differential equation rather than a simple linear one? Why a sinusoidal term to describe hypocrisy instead of random noise? Why an inverse-square field for divine attraction rather than a linear one? Here we explain the internal design choices of each model and how they align with the nuances of the verses and their exegetical tradition.

How This Explains the Relationship Best – Finally, we step back and show you what the model buys us. How does this formalization make the verse's structure more transparent? What does it allow us to compare, simulate, or generalize that a purely verbal description would struggle to express? Where does it illuminate the Qur'an's internal coherence or reveal hidden symmetries between passages?

By the end of this chapter, our hope is that you will not merely have *seen* mathematical models attached to familiar verses, but that you will begin to *think* in CRS terms yourself: identifying variables in a passage, discerning the type of relationship, and asking which mathematics best fits the Qur'anic logic at hand. You will see that the Qur'an, read through this lens, is not a passive object being forced into alien formalism. It is an active, deeply structured discourse that invites – and withstands – rigorous modeling, much as the natural and social worlds do. The difference is that here, the "system" we are modeling is not just physical or social, but revealed: an interface between the human mind and the divine will.

NOTE: The mathematical models presented for Surah al-Fāti ah (Surah 1) and the initial passages of Surah al-Baqarah (Surah 2) in this chapter are provided solely for illustration, offering readers a glimpse into the analytical approach used throughout this work. These examples demonstrate the modeling methodology rather than representing the complete structural, cognitive, and mathematical treatment of the verses. For the full models, comprehensive analyses, and the standardized process for developing them, please refer to the detailed guidelines in Chapter 9.

Crs Modeling Of Surah Al-Fātiḥah

Surah al-Fatihah is the first chapter of Quran.

1. Verses 1:2–4 – Divine Ontology & Praise Logic
   a) Variables

From UQVD and this unit's model, the key variables are:
Divine Attribute Variables (A_X):
A_R: Rahmah (mercy/compassion)
A_P: Qudrah (power/might)
A_J: 'Adl (justice)
A_K: 'Ilm (knowledge/omniscience)
A_U: Ahadiyyah (oneness/uniqueness)
A_S: amad (self-sufficiency/refuge)
These are collected into:
Divine Attribute Set:

$$D(t) = A_R, A_P, A_J, A_K, A_U, A_S$$

Divine Attribute Vector:

Praise\ Function:

b) Nature of the Relationship

Conceptually, the relationship is:
Human praise is a function of recognized divine attributes.
We are modeling a structural dependency:

c) Why this type of math? (Set Theory + Linear Algebra)

We use set theory and linear algebra (vectors) here because:

1. Enumeration and grouping: the verses enumerate multiple attributes, and set notation D(t) = {…} is ideal for capturing the collection.
2. Component-wise structure: a vector makes each attribute a component of a single unified Divine "state description."
3. Future extensibility: linear algebra allows norms, inner products with human state vectors, and matrix transformations later.

d) Why this specific model form (set + vector + function)?

We model:

1. Composite Divine Set: = signaling co-active attributes.
2. Vector Form: , enabling standardized operations.
3. Praise Function: =, because praise is a rational response to recognized attributes.

e) How this explains the relationship best

The vector-and-function representation:

- Captures simultaneity of divine qualities.
- Formalizes the intuition that praise rises as recognition deepens.
- Makes it easy to compare this attribute vector with other passages or later Biblical models.

2. Verse 1:5 – Servant–Lord Relationship
   a) Variables

From the Surah 1:5 unit:
Human-side state variables:

- : Islam (submission)

• : Iman (faith)
• : Taqwa (God-consciousness)
Devotion Vector:

Divine response variables:

• : Hidayah (guidance-giving capability)

• : Qudrah (power/might)

Response Vector:

Relationship Function (conceptual):

Vector level:

b) Nature of the Relationship

The relationship is a relational mapping from a human devotion state to a divine aid state.

It is modeled as a functional dependency: the human devotion vector feeds into a response function that outputs guidance and might.

c) Why this type of math? (Linear Algebra + Basic Functional Analysis)

1. Linear algebra (vectors) groups multiple aspects of devotion into a single "state" object and similarly for divine response.
2. A function mapping expresses the conditional nature of divine response to worship and reliance.

We avoid calculus here because the verse is about structural role alignment, not explicit time evolution.

d) Why this specific model form (devotion vector → response vector)?

Verse 1:5 defines roles:
• Human: worship + seeking help.
• Divine: guiding + empowering.
Therefore:
captures how pure/intense the human side is.
• expresses the two core channels of divine engagement.

• captures the idea that the quality of worship and reliance affects the quality and extent of divine aid.

e) How this explains the relationship best

It shows that "iyyaka naʿbudu wa iyyāka nastaʿīn" is a structural contract:
Input: aligned submission + faith + God-consciousness.
Output: guidance + empowerment.
It also aligns seamlessly with the previous unit where gives the capacity of God to respond and determines when and how that capacity is engaged.

3. Verses 1:6 –7 – Guidance Dynamics & Path Differentiation
   a) Variables

From this unit's model:
Divine-side guidance variables:
• : Hidayah (guidance-giving capability)
• : Hidayah Vector – the directional "force" of guidance.
Human-state variables:
• : Taqwa (God-consciousness)
• : Iman (faith)
• : Kufr (disbelief)
• : Nifāq (hypocrisy)
• : Doubt (rayb/shakk)
Human State Vector (example):
Deviation Set:

b) Nature of the Relationship

Guidance is modeled as a vector-valued function of human belief/taqwa, subject to exclusion from a deviation set.
Input: the internal state H (taqwa + iman).
Transformation: the guidance vector pointing along a certain path.
Constraint: , expressing that the requested path is not the path of those who incurred wrath or went astray.

c) Why this type of math? (Linear Algebra + Set Theory / Classification)

1. Linear algebra (vector guidance): guidance is inherently directional; a "Straight Path" is best represented as a trajectory or direction in state space.

2. Set theory (deviation set D): verse 7 provides two negative exemplars, which is naturally expressed as sets of states to avoid.

We avoid calculus here because the focus is on classification and request, not explicit time evolution.

d) Why this specific model form (guidance vector + deviation set + exclusion condition)?

Verse 1:6 is a vector request: "Guide us along the Straight Path," hence as a directional function of internal states.

Verse 1:7 introduces comparative classes (blessed vs. wrath vs. astray), which motivates set partitions.

The condition $G \notin D$ compactly encodes "not their path" as a formal constraint.

e) How this explains the relationship best

It shows that guidance is multi-dimensional and directional, consistent with the metaphor of a path.

It clarifies that verse 7 performs a filtering operation among all possible trajectories.

This framework also aligns the Surah's three segments:

• 1:1–4 defines who is guiding (attribute vector).

• 1:5 defines how we relate and receive help (devotion → aid mapping).

• 1:6–7 defines what we are specifically asking for in guidance (a constrained trajectory in moral space).

CRS Modeling of Surah 2 (Verses 1–20)

This document summarizes the Computational Religious Studies (CRS) modeling choices for the first thematic unit of Surah 2 (Al-Baqarah, Verses 1–20) as developed in the Quranic Modeling Project. The focus here is strictly on:

a) Variables
b) Nature of the relationship
c) Why a certain type of mathematics was used
d) Why the model took its specific mathematical form
e) How this explains the relationship best
   1. Surah 2:1–20 – Revelation and Belief Criteria

Thematic focus: Revelation as a doubtless Book and the differential response of three archetypes – believers, disbelievers, and hypocrites.

a) Variables

The model introduces a common belief-state framework with the following key variables:

• : Divine Guidance – time-dependent input from the Quran.

• : Scriptural input signal – the "Quranic stimulus" as it impacts the inner state.

• : Waswasa (temptation/distraction) – disturbing influences opposing .

Belief-state functions for different archetypes:

• : Belief state function for the faithful (mu'minūn).

• : Belief state function for the disbelievers (kuffār).

• : Belief state function for the hypocrites (munāfiqūn).

System parameters (psychological coefficients):

• : Damping coefficient (emotional/cognitive resistance or friction), indexed by i ∈ {I, K, N}.

• : Restorative or rigidity coefficient (how strongly the state tends to revert or resist change).

• : Angular frequency capturing repetitive, cyclical patterns of hypocrisy.

• : Noise term capturing random perturbations, rationalizations, or situational shocks.

• : Scaling constant for receptivity to scriptural input.

• : Forgetting or resistance decay rate in the integral kernel.

Core differential form (generic belief trajectory):

Specialized per archetype:

Believers :

Disbelievers :

, with for all .

Hypocrites :

This oscillates with phase instability and noise. Their behavior mimics belief without convergence:

Integral representation for cumulative guidance reception:

Where represents forgetting or resistance decay rate. This models spiritual receptivity weighted over time.

b) Nature of the Relationship

These equations express belief as a dynamic, time-evolving response to revelation:

1. Dynamic Input–State Relationship:

and act as external input signals (forces) into an internal belief system.

represents opposing forces (temptation, distraction, arrogance) that subtract from the effect of revelation.

2. Second-Order Internal Dynamics (Believers):

For , the second derivative captures acceleration in faith – how rapidly spiritual commitment increases or decreases in response to guidance.

The first derivative term reflects friction or resistance (e.g., doubts, laziness, social pressure) which slows or destabilizes spiritual growth.

The term reflects internal structure: existing habits, convictions, and cognitive schemas.

3. Inertial Null State (Disbelievers):

for all models complete epistemic closure: there is no internal motion or growth in response to revelation. The state is a flat line.

4. Oscillatory Instability (Hypocrites):

expresses ongoing fluctuation in the hypocritical state. There is no stable equilibrium; the system keeps oscillating between apparent commitment and withdrawal, with noise capturing situational shifts.

5. Memory and Forgetting (Integral G(t)):

represents guidance as a memory-weighted accumulation of past scriptural exposure: recent inputs count more (exponential kernel), older inputs fade due to forgetfulness (λ) or resistance.

c) Why This Type of Math (Calculus, Differential Equations, Signals, Integrals)?

1. Calculus and Second-Order Differential Equations for Believers:

The Qur'anic portrayal of believers in these verses is not static; it involves continuous prayer, charity, and responsiveness to unseen truths. This naturally lends itself to modeling as a dynamic system where the state of belief changes over time.

Second-order ODEs are standard in physics and control theory for systems with inertia and damping (e.g., mass-spring-damper). This captures both the momentum of spiritual growth and the friction that slows it.

2. First-Order Constant Equation for Disbelievers:

The statement that it is the same whether you warn them or not corresponds to a state that does not change with time or input. A simple first-order equation with is the most direct formalization of such epistemic sealing.

3. Sinusoidal Signal + Noise for Hypocrites:

Hypocrisy is characterized by internal oscillation: moments of apparent faith followed by relapse, driven by social incentives and fear of exposure. A sinusoidal term captures that periodic swing, while accounts for irregularities (changing circumstances, shocks, rationalizations).

Using signal-like behavior emphasizes that hypocrisy is not random noise alone; it has a structured, recurrent pattern.

4. Integral Operator for Guidance Reception:

The integral is a convolution-type expression used in systems theory to represent how inputs accumulate over time with decay.

This matches the idea that Quranic guidance is experienced as a history of recited verses, past reminders, and cumulative learning – but that their effect diminishes without reinforcement.

d) Why It Is Modeled in This Specific Form

1. Unified Framework for Three Archetypes:

By starting from a generic second-order equation and then specializing for believers, disbelievers, and hypocrites, the model presents all three within a single mathematical paradigm. This mirrors the Qur'an's juxtaposition of these archetypes in one passage.

2. Believers as a Damped, Driven System:

Choosing a mass-spring-damper–type structure (second-order ODE with damping and stiffness) allows us to represent:

– Initial shocks of revelation (as external forcing),

– Transient oscillations as the person adjusts,

– Eventual stabilization into a mature, steady state of faith.

This is more realistic and expressive than a simple linear growth model, because it captures struggle, relapse, and eventual convergence.

3. Disbelievers as an Equilibrium Fixed at Zero:

• Setting for all t directly formalizes the scriptural claim that warning docs not produce change. It is a deliberate choice to mirror the theological idea of a sealed heart as an absorbing state.

4. Hypocrites as Non-Convergent Oscillators:

The choice ensures that never settles into a stable limit; there is ongoing fluctuation. This matches the description of hypocrisy as inner turbulence and inconsistency.

Using a sinusoidal driver retains structured repetition (e.g., repeated patterns of lying, pledging and breaking promises), while $\varepsilon(t)$ leaves room for irregular behaviors.

5. Exponential Kernel in G(t):

The term ensures that recent Quranic input has stronger impact, while older inputs are exponentially down-weighted, reflecting human forgetting and the need for constant dhikr and tilāwah (remembrance, activity).

This choice aligns with well-known models in neuroscience and learning theory, making the CRS model intelligible to scientific audiences.

e) How This Explains the Relationship Best
   1. Capturing Differential Trajectories from a Common Revelation:

All three archetypes (believers, disbelievers, hypocrites) receive or are exposed to the same revelation signal , yet their internal trajectories diverge dramatically.

The model makes this explicit: the divergence is not in the input, but in the internal structure and parameters (, , initial conditions, and whether the system is even responsive).

2. Making Qur'anic Descriptions Dynamically Testable:

Verses describing stabilized believers, sealed disbelievers, and oscillating hypocrites are often treated qualitatively. Here, they are translated into distinct dynamical regimes:

– Stable, convergent trajectories (believers),
– Static, null trajectory (disbelievers),
– Non-convergent, oscillatory trajectory (hypocrites).

This invites simulation and scenario analysis, bringing CRS closer to computational psychology and control systems.

3. Integrating Memory, Repetition, and Decay:

The integral expression for aligns with repeated mention of reminders (dhikr), recitation, and forgetfulness in the Qur'an. It offers a natural handle to model how long-term exposure shapes belief.

4. Respecting Theological Content While Adding Formal Clarity:

The choice of second-order ODEs, constant solutions, and sinusoidal drivers does not claim to exhaust the spiritual reality; rather, it highlights distinct structural properties implied by the text.

It becomes clear, for example, that Qur'anic descriptions of hypocrisy are not simply moral labels but descriptions of a particular kind of unstable cognitive–moral dynamics.

5. Providing a Reusable Modeling Pattern:

This framework for Surah 2:1–20 sets a reusable pattern for later chapters: revelation as an input, belief as a dynamical state, and archetypes as different parameterizations or structures of the same underlying system.

In that sense, it "teaches" the reader how CRS will continue to reason about belief and guidance throughout the rest of the Qur'an.

CRS Modeling of Surah 2 (Verses 21–29) – Technical Notes

This document summarizes the Computational Religious Studies (CRS) modeling choices for Surah 2 (Al-Baqarah), Verses 21–29. The focus is on:

a) Variables
b) Nature of the relationship
c) Why a certain type of mathematics was used
d) Why the model took its specific mathematical form
e) How this explains the relationship best

Thematic Unit: Divine Authority and Scriptural Continuity (2:21–29)

The passage asserts divine authority through cosmological reference, rational challenge, and analogy. It calls humanity to worship the Creator, invokes signs in the earth and sky, issues an inimitability challenge to produce a surah like the Qur'an, and legitimizes parables (even of a mosquito) as vehicles for conveying abstract principles.

a) Variables

Mapped variables from the Universal Quranic Variable Dictionary (UQVD) and local modeling constructs:

Core divine and cognitive variables:

• : Qudrah (Power / Might) – divine power and creative authority.

• : 'Ilm (Knowledge) – divine knowledge underlying precise cosmological order.

• : Qur'an / Scriptural input – the revealed content as a signal or message.

• : Epistemic Filter – the human cognitive lens or filter that processes .

• : Submission Energy – spiritual potential or energy needed for an agent to attain submission.

• : Doubt Resistance – epistemic resistance or hardness toward revelation.

Spatial and field variables:

• : Spiritual space containing agents.

• : Position of an agent in spiritual space.

• : Center of divine ontological authority (reference point of the field).

• : Divine potential field at position .

• : Force exerted at position by the divine field.

• : Universal spiritual constant scaling the strength of divine attraction.

• : Unit vector pointing from toward .

Signal and filtering variables:

• : Quranic signal as a function of time t (recitation, reflection events).

• : Human response signal after filtering by .

• : Impulse response of the epistemic filter .

• : Integration variable (past time index).

Information-theoretic variables:

• : Class of Quranic content.

• : Human composition space (all possible human-authored texts).

• : A candidate human text proposed to imitate the Qur'an.

• Minimum Description Length complexity measure for a text.

Analogical reasoning variables:

• : Concrete symbol (e.g., mosquito, small object in a parable).

•: Abstract principle or moral lesson encoded by .

• : Analogical mapping function from symbol s to principle .

Key equations from the model:

1) Divine potential field:

Induced force:

2) Epistemic filtering of the Quranic signal:

3) Quranic inimitability via MDL:

This expresses that no human-produced text can match Quranic complexity with equal or shorter formal language encoding.

4) Submission energy differential:

Define the energy required to attain submission as:

Where ES > 0 implies net pull toward submission. If > , submission is repelled.

5) Analogical reasoning mapping:

Let be an analogical mapping function between concrete symbol and abstract principle p:

b) Nature of the Relationship

The unit's models capture several intertwined relationships:

1. Divine Authority as an Attractive Field:

The potential field and force formalize divine authority as a central attractive influence in spiritual space. Agents at different positions x experience varying intensities of inward pull toward (the locus of full submission and recognition of divine Lordship).

The relationship between and is direct: the stronger the recognized divine power, the stronger the cognitive/spiritual force urging worship.

2. Revelation as a Signal Processed by Human Cognition:

- represents the Qur'an as a time-varying input (recitation, encountered verses).
- expresses that the human response is not a direct copy of the input but a filtered version shaped by – the impulse response of the cognitive filter.
- (knowledge) and (doubt resistance) shape , so the same revelation yields different responses for different individuals.

3. Inimitability as a Complexity Inequality:

The MDL inequality expresses a structural separation between the Quranic content class and human composition space : no human text can achieve equal or lower description length while matching 's informational richness.

The relationship here is comparative: Quranic content sits at a lower MDL (more compressed, highly structured) than any imitative human text.

4. Submission Energy as Net Work of Divine Pull over Resistance:

models submission as dependent on the time-accumulated difference between divine pull () and resisting factors ().

If dominates over an interval, and the agent is pulled toward submission; if , the net effect is repulsion or stagnation.

5. Analogical Reasoning as Symbol-to-Principle Mapping:

says that lowly or simple symbols (like a mosquito) carry abstract principles when the analogical mapping is active.

The relationship is interpretive: the symbol itself is not the principle, but A enables receivers to transition from the concrete to the abstract.

c) Why This Type of Math (Fields, Calculus, Signals, Information Theory, Mappings)?
   1. Field Theory (Potential and Gradient):

The Qur'anic verses about creation of earth and sky, provision of rain, and growth of fruits suggest a pervasive, omnipresent influence of divine power on all aspects of life.

Potential fields and associated forces are standard tools in physics (e.g., gravity, electrostatics) for modeling central, distance-dependent influences. This makes them a natural fit for representing divine authority that attracts hearts from any position in spiritual space.

2. Calculus and Integration for Submission Energy:

Submission is not a single event but an accumulation of experiences, realizations, and internal struggles over time.

Using an integral mirrors the idea of work done by a force over an interval, capturing the cumulative effect of divine pull versus resistance.

3. Signal Processing (Convolution / Filtering):

The Qur'an is recited, heard, and pondered as a temporal sequence, making a signal perspective appropriate.

Convolution is the foundational operation in linear systems and signal processing, precisely modeling how an input is shaped by a system's impulse response.

It allows h(t) to encode sensitivity to certain "frequencies" or types of content (e.g., some hearts respond to rational arguments, others to parables, others resist both).

4. Information Theory and Minimum Description Length (MDL):

The inimitability challenge ("produce a surah like it") is naturally related to questions of information content, structure, and compressibility.

MDL provides an objective, formal way to compare complexity: texts that achieve more meaning with shorter optimal encodings are informationally superior.

Using MDL avoids subjective aesthetic criteria and ties the ijaz (miraculous nature) of the Qur'an to measurable structural properties.

5. Abstract Mapping for Analogical Reasoning:

Parables and examples are about mapping from concrete cases to general principles.

A simple mapping is mathematically minimal yet conceptually exact: it identifies analogy as a function from symbols to principles.

This prepares the ground for more advanced representations (e.g., graph-based or category-theoretic models of analogy) without overcomplicating the foundational exposition.

d) Why It Is Modeled in This Specific Form

1. Potential and Force:

The choice follows the structure of Newtonian gravity, where potential decreases with distance and its gradient yields an inverse-square force.

This specific form has interpretive advantages:

– Near (high proximity to God-consciousness), small shifts produce large changes in force, reflecting intense spiritual dynamics.

– Far from , the force weakens but never becomes zero, matching the Qur'anic idea that signs of God are still present everywhere.

2. Convolution-Based Epistemic Filter:

The form (with limits from to) is chosen because it explicitly models how past exposures to revelation continue to shape present response, weighted by h.

This avoids crude "one-shot" models (single verse → instant, permanent change) and instead reflects the Qur'an's emphasis on repeated reminders and gradual purification.

3. MDL Inequality as Inimitability Statement:

Using is the cleanest way to say: "Even the best possible human attempts are structurally and informationally inferior."

Other complexity measures (e.g., simple length or vocabulary size) can be gamed; MDL directly encodes the trade-off between model complexity and data fit, aligning with modern notions of scientific and linguistic elegance.

4. Submission Energy as Difference of Power and Resistance:

is defined as an integral of rather than, say, alone, because the passage itself juxtaposes divine creative signs with human doubt and contestation.

Making explicit allows the model to capture hardening of hearts, social pressures, and ego that push against the gravitational pull of revelation.

5. Simple Function for Analogical Reasoning:

The goal of this unit is to show that even small or lowly symbols can carry profound principles.

A linear, direct mapping from to foregrounds the essential logical structure without yet introducing more advanced machinery. It aligns tightly with the verse's claim that Allah is not ashamed to use a mosquito as an example.

e) How This Explains the Relationship Best

1. Unifying Cosmology, Epistemology, and Revelation:

The potential field and force situate divine authority in a spatial metaphor: everything in the cosmos experiences the pull of the Creator who made the earth a bed and the sky a canopy.

The signal-and-filter model then connects this cosmological authority to human epistemology: humans do not just exist in the field; they interpret its signs through cognitive filters.

The MDL-based inimitability framework shows that the revelation signal itself is uniquely structured, distinguishing it from any human attempt.

2. Making Resistance and Submission Quantifiable:

By defining as an integral over (), the model treats submission and resistance as quantifiable magnitudes rather than vague dispositions.

This allows comparative analysis: two individuals exposed to similar signs may accumulate very different submission energies depending on their resistance profiles.

3. Clarifying the Rational Structure of the Inimitability Challenge:

The MDL inequality reframes "produce a surah like it" as a challenge about structural and informational parsimony, not merely style or rhetoric.

It thereby emphasizes that ijaz is not just miraculous eloquence but also unmatched efficiency of encoding guidance, law, and narrative.

4. Elevating Parables from Decoration to Cognitive Mechanism:

Modeling analogy as underscores that parables (including those involving small creatures) are core mechanisms for encoding and transmitting principles.

This validates the Qur'an's explicit defense of parable usage: what some mock as trivial is in fact a mathematically recognizable mapping between domains.

5. Providing a Coherent CRS Lens on the Whole Unit:

Together, the models show that Verses 21–29 are not a random collection of exhortations, challenges, and examples. They form a logically coherent system:

– A central authority field grounded in creation.
– A unique revelation signal () with inimitable structure.
– Human filters () that shape response.
– A dynamic energy balance () capturing submission vs. resistance.
– An analogical engine () for lifting minds from concrete signs to abstract truths.

In CRS terms, this unit becomes a worked example of how cosmology, epistemology, and information structure are integrated in the Qur'anic discourse.

Thematic Unit: Adamic Narrative and Divine Delegation (Verses 2:30–39)

This part summarizes the Computational Religious Studies (CRS) modeling choices for Surah 2 (Al-Baqarah), Verses 30–39 (the Adamic narrative). The focus is on:

a) Variables
b) Nature of the relationship
c) Why a certain type of mathematics was used
d) Why the model took its specific mathematical form
e) How this explains the relationship best

Thematic Unit: Adamic Narrative – Symbolic Epistemology and Moral Control (2:30–39)

These verses present the appointment of Adam as khalifah (vicegerent), the teaching of names, the demonstration before the angels, the refusal of Iblis to prostrate, and the episode of error and repentance. The CRS model treats this as a tightly coupled system of symbolic cognition, epistemic ranking, and moral control with feedback via repentance.

a) Variables

The key variables and constructs used in the model are:

Symbolic cognition layer:

- Set of all entities whose names are taught to Adam.
- : Space of semantic symbols or meanings.
- : Semantic content or meaning corresponding to entity .
- , a mapping from entities to meanings (a name–meaning map).
- : Divine Knowledge – here realized as the complete function f known by Allah.
- : Human Agent – Adam, endowed with the acquired map (symbolic grounding).

Epistemic ranking layer (angels' beliefs):

• : Initial evidence state (before Adam's demonstration).

• : Updated evidence state (after Adam demonstrates knowledge of names).

• Probability assigned by the angels to the proposition "Adam (HA) is fit for khalifah-hood" given evidence .

• : Probability assigned to the belief that angels are more fit than Adam.

• : Prior probability of Adam's fitness.

• : Likelihood of observing evidence if Adam is indeed fit.

• : Marginal probability of observing.

Moral control and trajectory layer:

• : Moral state of the agent (Adam or Iblis) at time t.

• : Divine law (constraint set) governing morally valid trajectories.

• : Moral potential function; its gradient guides the direction of moral improvement.

• : Dirac delta impulse at time , modeling a sharp intervention (repentance event).

• : Moral trajectory of agent i parameterized by s over an interval .

• : Instantaneous moral reward or return for agent i along trajectory , possibly modulated by access to knowledge.

• : Utility or integrated moral return for agent over the trajectory.

• : Utility for Adam

• : Utility for Iblis.

Key equations from the model:

I. Symbolic Cognition Model:

Adam's symbolic grounding is expressed as:

II. Bayesian Epistemic Ranking:

Initial ordering of beliefs:

After knowledge demonstration, the posterior increases:

Bayes' update for Adam's fitness:

III. Moral Integrity Control Dynamics:

Moral state evolution under divine law LD:

Utility of a moral trajectory over :

Special conditions:

For Iblis:

For Adam:

is bounded

b) Nature of the Relationship

The relationships captured by these models are layered and interdependent:

1. Symbolic Knowledge as the Basis of Human Superiority:

The mapping represents divine bestowal of symbolic knowledge – the ability to link signs (names) to meanings ().

Adam's state "" indicates that he internalizes this map; his cognition is now structured by a complete symbolic system.

The relationship here is functional and structural: human status (khalifah potential) depends on mastery of an entire map from entities to meanings.

2. Angels' Epistemic Ranking Updated by Evidence:

Initially, angels hold , meaning they see themselves as better candidates.

After observing Adam's demonstration (), the belief that "Adam is fit" increases via Bayes' rule.

The relationship is probabilistic: new evidence reshapes the ranking between competing hypotheses about who is more fit.

3. Moral State Dynamics Under Divine Control:

expresses moral movement as a gradient descent on a potential function: the moral state is drawn toward minima that correspond to stable righteous configurations.

The term models a sudden corrective impulse (repentance) that instantaneously changes direction or magnitude of at time *T*.

4. Moral Trajectories and Utilities for Adam vs. Iblis:

Each agent's moral journey is a trajectory

evaluated via a utility functional .

For Iblis, signifies persistent deviation from divine law, driving to negative infinity (unbounded moral loss).

For Adam, says that his trajectory is mostly within divine law with repentance as a corrective event, keeping bounded and recoverable.

c) Why This Type of Math (Sets, Mappings, Probability, Differential Equations, Integrals)?

1. Set Theory and Mappings for Symbolic Cognition:

The core scriptural emphasis is that Adam is taught "the names of all things." This is essentially a mapping problem: associating each entity with a semantic representation.

Using sets N and Σ, and a function f: N → Σ, precisely captures this structure. It turns the narrative into a formal model of symbolic grounding, central to language and cognition.

2. Probability and Bayesian Update for Angelic Beliefs:

The angels' initial question ("Will You place in it those who will cause corruption?") and subsequent acceptance after the demonstration of knowledge clearly depict a change in belief under evidence.

Bayesian probability is the standard mathematical language for belief revision: prior → likelihood → posterior.

Using Bayes' rule bridges theological claims (divine knowledge justifying human appointment) with a formal epistemic mechanism.

3. Differential Equations and Gradient Systems for Moral Dynamics:

Moral life is inherently dynamic: agents move through states of obedience, temptation, error, and repentance.

A gradient system *d /dt* = – *V()* captures how moral states are pulled toward low-potential (stable) regions representing alignment with divine law.

This is analogous to physical systems descending an energy landscape, making the moral system intuitively accessible to readers familiar with dynamics.

4. Dirac Delta for Repentance Events:

Repentance in this narrative is not a slow drift but a sharp turn: Adam "receives words" and is immediately restored.

The Dirac delta (t) is a standard tool to represent instantaneous impulses in control theory. It allows the model to reflect sudden moral discontinuities (Tawbah) within a continuous-time framework.

5. Integral Functional for Moral Utility:

The evaluation of Adam's and Iblis's paths is not based on a single moment but on the entire trajectory over time.

mirrors economic and control-theoretic notions of cumulative reward.

This suits a scriptural narrative where what matters is not only error or success at a point but the overall lived path, especially with repentance in view.

d) Why It Is Modeled in This Specific Form

1. as a Total Map (Exhaustive Naming):

The model takes and assumes is total (every has a). This reflects the Qur'anic language of "all the names," not a partial or approximate list.

Making the map total emphasizes that Adam's cognitive edge is not a few facts but a fully grounded symbolic system, legitimizing his unique role as khalifah.

2. Explicit Bayesian Inequality and Update:

Starting with P(Khilafah |) < P(Angelic Supremacy) and then showing P(Khilafah |) > P(Khilafah |) makes the re-ranking explicit.

Using the standard Bayes formula gives a clear mechanism: evidence has much higher likelihood given "Adam is fit" than given "Adam is not fit," causing a posterior shift in his favor.

3. Gradient Descent Form for Moral Dynamics:

d /dt = − *V()* is chosen over a simpler linear ODE because it naturally accommodates multiple moral equilibria (local minima), barriers, and basins of attraction.

This allows the model to express the idea that some moral states are more stable (easier to remain in) and that climbing out of a bad state is like climbing uphill in the potential landscape.

4. Impulsive Term for Tawbah:

Rather than adjusting parameters gradually, the model adds an impulse that instantaneously alters at .

This reflects the narrative structure: error, then a decisive act of turning back, then acceptance. It also explains why Adam's endpoint differs radically from Iblis's despite both committing an initial transgression.

5. Divergent Utilities: :

Encoding Iblis's persistent refusal as expresses that a trajectory outside divine law with no repentance has unbounded negative moral cost.

By contrast, ⇒ bounded shows that inclusion of Tawbah in the trajectory changes the evaluation: even with error, the path remains evaluable and redeemable.

e) How This Explains the Relationship Best

1. Clarifying the Knowledge–Authority Nexus:

The symbolic cognition model shows that Adam's superiority is epistemic: he possesses a grounded, comprehensive map from entities to meanings.

The Bayesian layer demonstrates how this epistemic superiority rationally updates the angels' prior beliefs about fitness for khalifah-hood.

Together, they capture the Qur'anic claim that human authority on earth is justified by unique cognitive capacities endowed by God.

2. Distinguishing Epistemic Rejection from Moral Failure:

Iblis is modeled not as lacking information (he is present, sees the demonstration) but as choosing a trajectory outside .

This aligns with classical tafseer: his problem is arrogance and disobedience, not ignorance. The model's structure separates epistemic access (is shared) from moral response (different).

3. Capturing the Transformative Power of Repentance:

The impulse term embodies the scriptural idea that repentance can sharply change one's moral trajectory.

By keeping bounded in spite of error, the model explains why Adam's story is ultimately one of restored alignment rather than permanent fall, without minimizing the seriousness of the initial lapse.

4. Making the Narrative Dynamically and Computationally Rich:

The gradient system and utility functional open the door to simulating different moral trajectories under various potentials and knowledge profiles .

This moves the story from a flat narrative to a dynamic system whose behavior can be explored, compared, and generalized to other agents.

5. Providing a Template for Human Moral Cognition in CRS:

This unit becomes a prototype for modeling human agents in CRS: symbolically grounded, epistemically evaluated, and morally controlled via differential dynamics with impulsive feedback.

In the broader chapter, it shows how the Qur'an encodes a full stack: from language and concept formation, to belief revision, to control of behavior over time, with repentance as a designed feature of the system.

Taken together, the models in this chapter offer what is, to our knowledge, the first systematic, mathematically explicit walk through the Qur'an at this level of structural detail. By treating verses not only as sources of guidance and law but also as objects of formal analysis – with clearly defined variables, relationships, and chosen mathematical frameworks – we open a new way of *seeing* revelation. This methodology does more than translate verses into symbols; it exposes the underlying architecture of Qur'anic reasoning: how guidance functions like a vector field, how belief behaves as a dynamic system, how parables operate as mappings between domains, and how repentance acts as an impulse that can redirect an entire moral trajectory. Because the approach is disciplined and repeatable, it can be extended Surah by Surah, compared across scriptures, and connected to contemporary work in systems theory, cognitive science, and philosophy. In this sense, CRS does not replace classical scholarship – it adds a new analytic dimension to it, turning the Qur'an into a living laboratory where theological, ethical, and spiritual claims can be explored with the full precision of modern mathematics.

Chapter 9
Algorithmic Reasoning, AI Agents, and Simulations in Computational Quran

ABSTRACT

Chapter 9 demonstrates how the Computational Qur'an (CQ) functions as a practical computational substrate for AI agents, simulations, and governance analysis rather than a purely theoretical framework. It shows how Qur'an-derived models, variables, and lemmas can be embedded into algorithmic reasoning systems that assist with ethical evaluation, decision support, and policy analysis while remaining transparent and inspectable. Through examples ranging from finance ethics and AI alignment to social simulations, legislation review, and scholarly advisory tools, the chapter illustrates CQ's role as a durable model layer beneath evolving AI architectures. It further explores data-driven calibration, lemma retrieval at scale, and human-in-the-loop learning, presenting CQ as a living system where revelation provides structural form and computation enables disciplined exploration across personal, institutional, and civilizational domains.

INTRODUCTION

In this chapter, we demonstrate how the Computational Qur'an (CQ) can be *used* – not just admired as a theoretical construct. We will walk through concrete applications that range from deeply personal questions to institutional policy analysis, and from system-level simulations to AI agents that reason with Qur'anic structures. The mathematical models you have seen – differential equations, logical systems,

DOI: 10.4018/979-8-3373-6940-2.ch009

set-theoretic constructions, and dynamic variables – are not tied to a single domain; they can be applied wherever ethics, behavior, and outcomes must be understood together. Nor are they confined to one generation or type of AI model: CQ can sit beneath today's large language models, retrieval systems, and decision-support tools, and as newer paradigms such as nested learning architectures, world models, and even quantum algorithms are formalized and adopted, the same Qur'an-derived structures can be mapped into those frameworks as well. In other words, CQ is designed as a model layer that endures while the AI stack evolves around it.

1. ALGORITHMIC REASONING AND AI AGENTS

Up to this point, we have treated the Computational Qur'an primarily as a modeling framework – a way to represent verses as systems of variables and relationships. In practice, however, those systems do not need to remain static diagrams. They can be embedded inside algorithms and AI agents that reason, evaluate, and explain in ways that are explicitly constrained by Qur'anic structures.

In this section, we present two concrete examples:

1. A CQ-governed question–answering (QA) agent that assists a finance ethics board in evaluating a new loan structure.
2. An "ethical alignment checker" that evaluates the behavior of AI systems – such as content moderators – against Qur'anic criteria for just and dignified speech.

Both examples follow the same general pattern:

- Encode a real-world situation into the CQ variable space.
- Retrieve relevant lemmas (local theorems) from CQ that capture Qur'anic conditions.
- Apply those lemmas as constraints or evaluators over the encoded situation.
- Use an AI component to aggregate results and produce explanations, while keeping the underlying logic anchored in CQ models.

These are not fatwā engines or substitutes for scholarly judgment. They are structured assistants: they help organize the relevant Qur'anic logic, make it computable, and present it in a transparent way.

Example 1: A CQ-Governed QA Agent for Justice in Finance

Imagine a Muslim finance board, or an institutional ethics committee, facing a very practical question:

"Is this new 'income-sharing loan' structure consistent with Qur'anic principles of justice and avoidance of ribā?"

The proposal may stretch across tens of pages, full of clauses, contingencies, and legal language. Traditionally, scholars would read it, recall relevant verses and fiqh, and work through the implications. With CQ, that process can be structured algorithmically without replacing the scholars themselves.

CQ Ingredients

The first layer is the variable space. The CQ models on finance, trade, and social justice already introduce variables such as:

W_D: wealth distribution (before and after the contract).

R_R: risk responsibility – how much risk is borne by each party.

RIBA: interest-like exploitation index – how close the structure is to a guaranteed, risk-free gain.

J_S: justice score – an aggregate indicator derived from distribution, symmetry, and opportunity.

O_P: oppression pressure – how likely the structure is to place one party under undue burden.

Across the Qur'an, the Computational Qur'an corpus includes lemmas related to:

- Prohibition of ribā (e.g., models derived from 2:275–279).
- Encouragement of charity, debt relief, and fair trade.
- Warnings against oppression and power asymmetry in contracts.

In CQ, a lemma is a small modeling unit – essentially a mini-theorem – that combines:

- A formal condition (for example: "if return is guaranteed and risk is not shared, then the exploitation index increases above threshold").
- Scriptural grounding (which verses this condition is anchored in).
- A model fragment (inequalities, functional forms, or logical constraints).

These lemmas are the building blocks of algorithmic reasoning.

Step 1: Policy Encoding Layer

The first step is a symbolic front-end that converts the human contract into CQ variables. This is not magic; it is an explicit mapping process, possibly assisted by natural language processing but always inspectable.

Key questions include:

- How is risk apportioned between lender and borrower? (Maps to R_R).
- Is there any guaranteed, risk-free return, regardless of outcome? (Influences *RIBA*).
- What happens in default or hardship? Are there provisions for relief or is the burden strictly enforced? (Affects O_P and post-contract W_D).

The result is an abstract representation such as:

$$Contract = R_R, RIBA, W_{D_{pre}}, W_{D_{post}}, G_R, \ldots$$

where G_R might represent "guaranteed return strength," and so on.

This encoding is crucial: it makes the contract visible to CQ models in a structured way.

Step 2: CQ Lemma Retrieval – The "Semantic Engine"

Once the contract is encoded, the system must determine which parts of the Computational Qur'an are relevant.

To do this, we use embeddings keyed to the UQVD. Each lemma in the CQ corpus is tagged with:

- Variables (e.g., $RIBA, W_D, O_P$)
- Thematic labels (finance, charity, trade, oppression, justice).
- Mathematical structure.

We then retrieve a lemma set:

$$L_{finance} = \ell 1, \ell 2, \ldots, \ell n$$

where each ℓ_i is a finance-relevant lemma: perhaps a condition on ribā, a model of fair profit-sharing, or a pattern relating unjust enrichment to social destabilization.

This is, in effect, a semantic engine: given "income-sharing loan," it finds all CQ lemmas whose variables and tags intersect meaningfully with the encoded contract.

Step 3: Constraint Checking and Rule Aggregation

With both the contract and relevant lemmas in hand, the system can now evaluate. For each lemma in :

- Plug the contract variables into the lemma's condition.
- Determine whether the lemma is:
  - Clearly satisfied.
  - Clearly violated.
  - Ambiguous or partially satisfied (e.g., depends on contextual thresholds).

Motivated by this, the system computes an aggregate evaluation vector, for example:

Here

summarizes justice-related lemmas.

indicates whether any ribā-like structure seems present.

captures oppression pressure.

measures how closely the contract resembles Qur'anic ideals of relief and support.

A simple rule-based layer or a small neural classifier can then map S into high-level categories, such as:

- Strongly aligned (most lemmas satisfied; no major red flags).
- Conditionally aligned (alignment depends on safeguards or implementation details).
- Misaligned (core lemmas about justice or ribā are violated).

This is not the final judgment; it is a structured answer to the question, "How does this contract look when viewed through the lens of CQ models?"

Step 4: Explanation Generation

The final step is explanation, where modern AI tools are particularly helpful. A large language model (LLM) is used here, but with strict constraints:

- Its knowledge base is limited to CQ artifacts: the lemmas, equations, and explanatory notes already curated by scholars and modelers.
- Its task is to summarize and explain, not to invent new rulings.
- Given the evaluation vector and the list of lemmas triggered, the LLM produces:
- A human-readable explanation of which lemmas were activated and why.
- Identification of conditions that were satisfied (e.g., risk-sharing is robust, no guaranteed interest).
- Identification of conditions that failed or are ambiguous (e.g., harsh default penalties raising oppression risk).
- Suggestions for modifying the contract to improve alignment: for instance, adding hardship relief clauses or adjusting profit-sharing ratios.
- The output is something a finance board can read, critique, and discuss. The system does not issue a fatwā; it prepares a disciplined, CQ-based analytic frame to support the board's deliberation.

What This Example Demonstrates

This example shows how the Computational Qur'an can be embedded into a hybrid AI system:

- Retrieval of topic-relevant lemmas.
- Symbolic evaluation of a real-world contract against formal CQ conditions.
- LLM-based explanation grounded in an existing CQ corpus.

It also illustrates how lemmas scattered across the entire CQ on a single topic – ribā, justice, wealth distribution – can be assembled algorithmically to answer a concrete question. Importantly, the system's reasoning path is inspectable: scholars can see which lemmas were used, how the contract was encoded, and where the alignment scores came from.

Example 2: An "Ethical Alignment Checker" for AI Systems

The same pattern can be applied outside finance. Consider an AI lab that has built a content-moderation system and now asks:

"Does this system respect Qur'anic principles of justice, dignity, and avoidance of slander and backbiting?"

Instead of evaluating only with standard technical metrics, we can add a CQ-based ethical lens.

CQ Ingredients

From the CQ corpus, we draw variables such as:

: honor/reputation level of individuals or groups affected by content.

: slander rate – frequency of false or harmful statements.

: gossip/backbiting index – how often content fits Qur'anic descriptions of backbiting.

: fairness of judgment – whether decisions are applied consistently and without bias.

: evidence strength – how strongly claims are supported by reliable information.

We also have lemmas derived from verses on:

Gossip and backbiting (for example, 49:12), modeled as negative externality functions: content that harms honor without benefit or justification.

False accusation and evidentiary thresholds: conditions under which serious allegations are forbidden without clear proof.

Justice and equity in speech: commands to verify information, avoid ridicule, and speak with fairness.

Each lemma encodes a simple conditional rule, such as:

- "If content causes a drop in without sufficient , it is unjust."
- "If content significantly increases with no countervailing benefit, it is harmful."

Step 1: Mapping System Outputs into CQ Variable Space

The moderation system produces outputs: approve, block, warn, down-rank, and so on, applied to specific pieces of content.

To evaluate these outputs, we:

1. Create or use a curated test set of example cases, possibly including synthetic scenarios that stress the system.

2. For each case, estimate the corresponding CQ variables:

- How severely does the content harm reputation? (approximate impact).

- Is there verifiable evidence behind the claims? (estimate).
- Does it fit patterns of gossip or backbiting? (estimate).

3. Combine these with the system's decision: for example, "high gossip index, low evidence, but content was allowed."

This mapping step can involve human labeling, auxiliary classifiers, or both, but the important point is that each case is now represented in CQ's ethical variable space.

Step 2: Applying CQ Lemmas as Constraints

Next, we apply CQ lemmas as constraints or evaluators over each case.

Consider a lemma of the form:

"Content that causes to drop below a threshold without sufficient should not be amplified (and ideally should be removed or countered)."

For each test case, we evaluate whether the system's decision:

- Satisfies the lemma (e.g., harmful, unsupported content was blocked or down-ranked).
- Violates the lemma (harmful, unsupported content was allowed or amplified).
- Is ambiguous (because variables are near thresholds).

Other lemmas might check whether:

- Non-harmful content was wrongly suppressed (unjust censorship).
- Genuine warnings (with strong evidence) were appropriately allowed.

By running all relevant lemmas on all cases, we can compute a CQ-Alignment score over the test set. This might be as simple as:

CQ_Alignment = 1 - (Number of serious lemma violations / Number of relevant cases)

or a more granular vector indicating performance on different ethical dimensions (honor protection, justice, evidence respect).

Step 3: Using CQ Alignment as a Performance Metric

Once computed, this CQ-alignment score becomes an additional metric alongside accuracy, precision, recall, or F1:

- Technical metrics tell us how closely the system's outputs match human-labeled "ground truth."
- CQ metrics tell us how closely its behavior aligns with Qur'anic ethical structures, as encoded in CQ.

Developers can use these metrics to:

- Compare versions of the system: "Does our new model improve CQ alignment?"
- Identify failure patterns: for example, repeated tolerance of gossip-like content, or over-suppression of legitimate criticism.
- Guide further training and policy adjustments.

Again, an explanation layer can summarize:

- Which CQ lemmas were violated most frequently.
- Which kinds of content tend to cause ethical misalignment.
- Concrete suggestions for modifying moderation policies or thresholds.

What This Example Demonstrates

This second example shows CQ functioning as a normative evaluation layer for AI systems:

- It does not generate content or make decisions by itself.
- It assesses the decisions of another system against Qur'anic criteria, as formalized in CQ.

By making these criteria explicit and computable, we move from vague appeals to "Islamic ethics" to a disciplined, model-based assessment. This does not replace the need for human oversight; rather, it provides a structured, repeatable way to check whether complex AI systems are drifting away from or converging toward Qur'anic principles.

Taken together, these two examples illustrate the broader point of this section: the Computational Qur'an is not only a theoretical framework; it is a computational substrate that can inform AI agents, decision-support tools, and evaluative systems. In each case, CQ provides the variables, lemmas, and models. Algorithms then operate on these structures – never replacing revelation, but making its internal logic accessible, testable, and explorable in an age defined by computation.

2. SIMULATIONS AND SYSTEM DYNAMICS

If the first class of applications showed how CQ can guide discrete judgments (Is this contract just? Is this AI system aligned?), simulations and system dynamics show something different: how Qur'anic logic behaves over time.

The Qur'an repeatedly links patterns of behavior to long-run consequences: oppression leading to collapse, gratitude leading to increase, heedlessness leading to decay, repentance reopening paths to renewal. CQ captures these patterns as dynamic models – systems in which variables evolve over time according to equations that encode Qur'anic relationships. Once those models exist, they can be simulated under different assumptions and policy choices.

In this section, we present two examples of such simulations:

1. A justice vs. oppression simulation for evaluating land-reform and zakat policies.
2. A belief trajectory simulation for youth education programs, exploring how different inputs affect long-term faith resilience.

In both cases, CQ provides the structure; simulation provides a way to "play out" Qur'anic dynamics under real-world assumptions.

Example 3: Justice vs. Oppression Simulation for Policy Design

Consider a government task force or advisory council faced with a familiar question:

"If we implement this land-reform and zakat policy, what happens to long-term social stability in light of Qur'anic principles?"

Typically, such a question is answered through economic analysis, political forecasting, and qualitative ethical reasoning. CQ does not replace those tools, but it adds an additional, distinctive layer: a dynamic justice–oppression model anchored in the Qur'an.

CQ Ingredients

Within CQ, we already have models describing:

- Distribution of wealth: how zakat, charity, prohibitions on hoarding, and injunctions against exploitation affect the flow of resources.
- Oppression vs. justice as dynamic systems: how increasing injustice feeds social tension, how reform can reduce pressure, and how delayed correction can lead to sudden breakdown.

A simple generic structure might look like:

where:

-OP is an oppression index, capturing the combined effect of unfair laws, abusive practices, and structural inequities.

- is wealth distribution (often summarized by a Gini-like measure in the real world).

- is power imbalance between social groups or classes.

- is enforcement of Qur'anic norms – how strongly the system enforces fairness, protects rights, and punishes oppression.

- represents justice institutions (courts, oversight bodies, accountability mechanisms).

The function f captures core Qur'anic intuitions: for example, that severe concentration of wealth and power combined with weak enforcement of justice will cause oppression pressure to rise, whereas redistribution and sincere reform reduce it over time.

Additional equations track related variables such as:

- : social cohesion.
- : trust in institutions.

These may depend on , and variables representing transparency, consultation, and responsiveness.

Step 1: Defining Policy Scenarios

To use this model in practice, the task force begins by defining scenarios:

- Scenario 1 – Baseline: no reform; current land distribution and zakat enforcement remain as they are.
- Scenario 2 – Moderate Reform: some land redistribution, improved zakat collection, and modest strengthening of justice institutions.
- Scenario 3 – Aggressive Reform: substantial redistribution, rigorous enforcement of zakat and anti-hoarding rules, and strong accountability mechanisms.

Each scenario corresponds to different parameter choices:

- How much does improve over time?
- How does change (weak, moderate, strong enforcement)?
- What is the effect on as large estates are broken up or regulated?

Step 2: Parameterization with Real Data

—— To move from abstraction to policy relevance, the model must be anchored in data. Here, conventional statistics and CQ modeling meet.

For each scenario, parameters and initial conditions are set using:

—— Gini coefficients or similar measures for wealth distribution.
—— Enforcement indices (for example, how often land laws are actually applied, or how consistently zakat is collected).
—— Complaint statistics, corruption indices, or survey data that approximate existing levels of oppression and trust.

Where exact data is unavailable, the model uses plausible ranges. Learning algorithms can be used to calibrate these parameters against historical patterns, but even without advanced fitting, reasonable ranges can still support comparative analysis.

Step 3: Running Time-Based Simulations

With scenarios and parameters in place, the system runs time-based simulations:

—— Integrate the differential equations forward in time – perhaps 10, 20, or 30 years.
—— Track trajectories for:

: oppression index.
: social cohesion.
: trust in institutions.

Graphically, one might see curves showing oppression pressure decaying under one scenario and remaining high or oscillatory under another.

For instance:

—— Under Scenario 1 (Baseline), the model might show staying high or rising, with declining and eroding.
—— Under Scenario 2 (Moderate Reform), may gradually fall below a "danger threshold" within 15–20 years, while) and stabilize and then rise.

- Under Scenario 3 (Aggressive Reform), the model might indicate a faster drop in oppression but also short-term volatility – perhaps reflecting social friction around rapid change – before achieving a more just equilibrium.

The point is not to pretend that a single simulation can predict the future with precision. Rather, it shows relative dynamics under different choices, guided by a Qur'an-structured model.

Step 4: Interpreting Results Through a Qur'anic Lens

Once the trajectories are generated, the interpretive work begins. CQ encourages reading the graphs in light of the verses that inspired the model.

For example, if persistent high oppression in Scenario 1 leads to eventual loss of cohesion and trust, this echoes repeated Qur'anic narratives: societies that tolerate injustice eventually face internal breakdown or external punishment. Conversely, trajectories in Scenario 2 that show oppression decaying and cohesion rising embody Qur'anic themes of reform (islāh) and fulfillment of rights leading to stability and blessing.

The task force can then frame its discussions along lines such as:

- "If we remain in a baseline state, our modeled oppression index never falls below the threshold associated with long-run stability. This resonates with Qur'anic warnings about unjust nations."
- "A moderate, consistent reform program seems to align both with the models and with the Qur'anic preference for sustained justice and relief of the oppressed."

Charts and simulations, in this context, are not merely technical tools; they are visualizations of Qur'anic logic applied to a particular policy question.

What This Example Demonstrates

This example shows that CQ models can generate policy-relevant simulations that go beyond rhetorical appeals. Rather than saying "Islam calls for justice" in the abstract, policymakers can see how different choices move a system through a Qur'an-informed state space over time. The model does not decide policy, but it provides a disciplined way to think about long-term consequences in light of Qur'anic structures.

Example 4: Belief Trajectories in Youth Education Programs

The same dynamic approach can be applied at a more personal, but equally important, scale: faith resilience in young people.

Imagine an educational NGO asking:

"What is the likely long-term impact on faith resilience if we increase high-quality Qur'an education by X and mentorship exposure by Y?"

Here, the unit is not a nation but a cohort of students, and the variable of interest is not oppression but faith intensity.

CQ Ingredients

CQ includes models of belief that reflect multiple Qur'anic themes:

- Reminders strengthening faith.
- Temptations and harmful environments weakening it.
- Social support and righteous company acting as protective factors.
- Harmful inputs – doubts, mockery, misinformation – eroding conviction if left unaddressed.

A simple dynamic model might be:

where:

- I(t) is faith intensity at time t.
- is the quality and quantity of Qur'anic reminders (education, recitation, reflective study).
- is temptation pressure (exposure to influences that pull away from faith).
- is supportive social environment (mentors, family, community, righteous peers).
-) is doubts and harmful inputs (unanswered questions, ridicule, hostile environments).
- The coefficients reflect how strongly each factor influences the trajectory.

This structure is not arbitrary; it reflects Qur'anic depictions of hearts being strengthened or weakened by what they hear, see, remember, and whom they accompany.

Step 1: Building a Cohort-Based Simulation

The NGO begins by conceptualizing its beneficiaries as a cohort of agents. Each agent (student) is assigned:

- An initial faith intensity I(0), drawn from a distribution (some begin strong, some weak, some in between).
- Exposure patterns describing typical levels of over time, based on current programs and broader environment.

The NGO's proposed intervention – say, a new mentorship scheme plus improved Qur'an curriculum – is represented as changes in and :

- increases due to better quality and regularity of Qur'anic education.
- increases as more students gain access to supportive mentors and peer circles.

Temptation pressure and harmful inputs might remain similar in the short term, reflecting wider societal conditions.

Step 2: Simulating With and Without the Program

Next, the NGO runs multi-year simulations of the cohort's faith trajectories under two conditions:

- Baseline: no new program; each student's evolves under existing exposure patterns.
- Intervention: the program is implemented; and are increased according to realistic assumptions (for example, additional weekly sessions, improved content, sustained mentorship).

For each condition, the system integrates the differential equation forward for each student, producing curves over, say, 5–10 years.

The NGO then examines key indicators, such as:

—— The probability that stays above a certain threshold (representing resilient, practicing faith).

—— The fraction of students whose trajectories show serious decline (dropping below another threshold).

—— How quickly the average faith intensity recovers after periods of elevated temptation or harmful exposure.

Step 3: Interpreting and Using the Results

The simulations might show, for example, that:

- Without the program, a significant portion of the cohort's faith trajectories drift downward over time, especially under increasing temptation pressure around certain life stages.
- With the program, the proportion of trajectories that remain above the resilience threshold is substantially higher, and even those that dip are more likely to recover as mentors and reminders act as stabilizing inputs.

For the NGO, this provides more than anecdotal reassurance. It offers a structured, Qur'an-informed way to quantify potential impact, enabling them to:

- Prioritize interventions with the greatest modeled effect on long-term resilience.
- Justify resource allocation to donors and partners with a model that explicitly reflects Qur'anic logic about remembrance, company, and protection from harmful influences.
- Design monitoring indicators (for example, proxies for) that are aligned with the underlying model.

At a deeper level, the simulation invites reflection on how youth programs relate to Qur'anic conceptions of hearts, guidance, and misguidance. The equations do not reduce faith to a number; they provide a lens for thinking about which levers – education, environment, mentorship – are most critical over time.

What This Example Demonstrates

This second example shows how CQ can inform program design and evaluation. Instead of relying only on short-term metrics (attendance, test scores), organizations can use CQ-derived models to reason about long-term spiritual trajectories.

The key insight is that the Qur'an already describes belief, reminders, environment, and temptation in dynamic terms. CQ turns those descriptions into mathematical systems, and simulation turns those systems into policy and program tools. Neither replaces the need for human judgment or spiritual care, but both provide a disciplined, transparent way to ask: If we act this way, over time, what kind of spiritual ecosystem are we building?

Taken together, the justice–oppression simulation and the belief trajectory simulation illustrate the power of treating the Computational Qur'an not as a static catalogue of rules, but as a dynamic system. When that system is made explicit and simulated, Qur'anic patterns of cause and effect become visible in a form that policymakers, educators, and community leaders can use – without ever losing sight of the revelation from which those patterns are drawn.

3. POLICY AND GOVERNANCE ANALYSIS WITH THE COMPUTATIONAL QUR'AN

Up to now, our applications have focused on how the Computational Qur'an (CQ) can guide discrete judgments and dynamic simulations. A third, equally important frontier is policy and governance analysis: laws, regulations, fatāwā, and institutional decisions that shape entire societies. In this context, CQ does not function as a replacement for parliaments, shūrā councils, or scholarly bodies. Instead, it provides a structured analytical layer that helps these bodies reason more transparently, systematically, and in explicit alignment with Qur'anic models.

Policy questions are often multi-dimensional: a media law affects freedom of expression, protection of honor, corruption dynamics, misinformation, and social cohesion – all at once. Scholarly questions about new biomedical procedures or climate agreements span multiple ethical themes: harm minimization, stewardship, justice between generations, and mercy. The traditional tools for handling such complexity – textual recall, analogical reasoning, and general ethical maxims – remain indispensable, but they are stretched by the scale and speed of contemporary problems.

CQ offers a way to encode Qur'anic ethical structures in a form that can be combined with modern methods: multi-criteria decision analysis, impact assessment, and graph-based reasoning. In this section, we illustrate this with two examples:

1. A CQ-based "ethical impact assessment" for proposed legislation, using a draft media regulation law as a test case.
2. A fatwā / advisory support tool that helps scholarly councils organize and test complex arguments by building reasoning graphs from the CQ lemma library.

In each case, CQ acts as a scaffold: it provides clearly defined variables, lemmas, and relationships. Algorithms then use that scaffold to analyze options and present structured views of their ethical implications. Human decision-makers – legislators or scholars – remain fully in charge of judgment and authority.

Example 5: CQ-Based Ethical Impact Assessment for Legislation

Consider a parliament or shūrā council that wishes to evaluate a draft law on media regulation. The law may include rules on platform liability, penalties for defamation, content takedowns, and responsibilities around misinformation. Traditionally, the debate will feature high-level ethical slogans – "protect freedom of speech," "protect honor," "prevent corruption" – but often lacks a consistent, formal framework that ties those themes back to Qur'anic structures in a measurable way.

A CQ-based "ethical impact assessment" introduces such a framework. Rather than treating the law as a collection of disconnected clauses, CQ encourages the council to see it as a vector of levers operating on a Qur'an-modeled ethical system. The assessment then becomes a matter of asking: when these levers move, how do they shift key variables such as harm to reputation, gossip, slander, unjust censorship, and corruption risk?

CQ Ingredients

At the foundation of this approach are lemma clusters from the Computational Qur'an that address speech, evidence, and justice, including models derived from verses on:

- Speaking with justice, avoiding mockery and ridicule.
- Prohibitions of slander, backbiting, and spreading corruption.
- Checking sources, avoiding haste in transmitting information.
- Standing for truth even when it is unpopular or costly.

In CQ, these themes are encoded through variables such as:

- : freedom of speech, understood as the system's capacity to allow truthful, beneficial speech to circulate.
- : harm to reputation (honor index), reflecting how often individuals' or groups' honor is damaged without just cause.
- : misinformation circulation, capturing the rate at which unverified or false content spreads.
- : corruption risk, representing the likelihood that speech norms enable or suppress corrupt behavior.

Each lemma specifies conditions involving these variables. For example, one lemma might state that "content that significantly reduces without sufficient evidence should be minimized," while another may encode the duty to verify information before passing it on, thereby constraining . Others express concerns about unjust silencing, ensuring that efforts to reduce harm do not slide into oppression or propaganda.

Representing the Law as a Vector of Legal Levers

The first algorithmic step is to represent the proposed law as a vector of legal levers:

Here:

- captures the range and severity of penalties for infractions such as defamation, incitement, or misinformation.

—— encodes evidentiary thresholds (for example, how much proof is required before sanctions are applied).
—— reflects what content platforms are required to do (monitor, report, remove, down-rank, notify users, etc.).
—— Additional components might capture appeals mechanisms, transparency obligations, and protections for whistleblowers.

This vector L is a compact description of the legal design space. Any revision to the law can be thought of as a change in one or more coordinates of L, making the structure amenable to systematic analysis and comparison between drafts.

Impact Functions from Law to CQ Variables

The next layer consists of impact functions derived from CQ models. Each legal lever in is mapped to its expected effect on the CQ variable space (and others relevant to the case). For example:

- Increasing may reduce (by deterring reckless spread of harmful content), but if applied without proper safeguards, it may also reduce and increase (if powerful actors use the law to silence critics).
- Tightening before punishment can protect
  by reducing false accusations, but if thresholds are too high, they may fail to deter genuine harmful content.
- Expanding may reduce and corruption-enabling discourse but may also create incentives for platforms to over-remove speech, harming .

These relationships are not invented ad hoc; they are encoded based on lemma clusters that specify how certain practices affect honor, justice, corruption, and speech norms in Qur'anic narratives and commands. The result is a set of functions such as:

which together define how a given legislative configuration L moves the ethical state of the system.

Multi-Criteria Evaluation and Qur'anic Alignment Index

Given these impact functions, the assessment proceeds with a multi-criteria evaluation. The council defines an index of "Qur'anic alignment" that combines key objectives with appropriate weights. For instance:

where:

—— is inversely related to (less unjust harm means higher protection).
—— reflects a combination of low and fair evidentiary standards.
—— measures success in reducing gossip and backbiting without suppressing justified criticism.
—— reflects safeguarding against unnecessary or oppressive restrictions.

The weights can be set in consultation with scholars and stakeholders, reflecting their understanding of priority among these objectives in the particular context. For a given draft , the system computes and also reports the underlying components, so that council members can see not just a single score but the structure behind it.

Sensitivity and Scenario Analysis

Because real-world implementation is uncertain, the CQ-based assessment can include sensitivity analysis. For example:

- How robust is to variations in enforcement intensity? If enforcement is weak, does the law still protect honor and reduce misinformation, or does it become a tool selectively applied against critics?
- How does change if social media usage increases dramatically, or if new forms of misinformation arise?

The system can simulate alternative enforcement scenarios by adjusting variables such as effective or actual platform compliance, then recomputing and the overall index. This allows legislators to see where the law is fragile and where it is robust, all within a framework explicitly grounded in Qur'anic models of speech and justice.

What This Example Shows

This example demonstrates CQ as a foundation for multi-criteria decision analysis in governance. Instead of relying on ad-hoc ethical slogans, the assessment is built on a network of formal lemmas that encode Qur'anic guidance on speech, honor, and corruption. The law is represented as a set of tunable parameters; CQ impact functions map those parameters to ethical variables; and a transparent alignment index summarizes how well the law performs across multiple Qur'anic criteria.

The result is not an automated verdict, but a disciplined way for human councils to see trade-offs, interrogate assumptions, and refine legislation in a manner that is both technically and scripturally coherent.

Example 6: Fatwā and Advisory Support Tool

Beyond formal legislation, many of the most consequential decisions in Muslim societies emerge through fatāwā, advisory opinions, and policy guidance issued by councils of scholars. These decisions increasingly involve topics that are cross-disciplinary and technically demanding: climate agreements, AI deployment, novel biomedical interventions, complex financial instruments, and more. In such cases, scholars must integrate a wide range of textual evidence, ethical principles, and domain-specific information.

A CQ-based fatwā / advisory support tool is designed to function as a cognitive exoskeleton: it assists scholars in organizing, exploring, and testing their reasoning without replacing their judgment or authority. Its core components are the CQ lemma library and a reasoning graph engine that can assemble these lemmas into candidate argument chains.

CQ Lemma Library and Tagging

Within CQ, lemmas are tagged not only by variables and equations, but also by richer metadata, such as:

- Ethical theme: justice, mercy, harm minimization, stewardship, trust, accountability, etc.
- Domain tags: environment, family, commerce, warfare, governance, health, technology.
- Scriptural grounding: specific sūrahs and verses, including cross-references to classical tafsīr where needed.

For example, lemmas relevant to climate policy may be tagged under themes like stewardship and corruption on earth, with domain tags such as environment and intergenerational justice. Lemmas relevant to new biomedical procedures might combine stewardship of the body, preservation of life, and avoidance of harm and injustice in medical practice.

Tool Behavior: From Question to Reasoning Graph

The basic workflow for scholars using the tool could be:

1. Question input: A council member enters a structured description of the problem – for instance, "A proposed international agreement requires our country to commit to certain emissions reductions and financial transfers; does endorsing this agreement align with Qur'anic principles of justice and stewardship?"

2. Lemma retrieval: The system searches the CQ lemma library using the question's ethical and domain descriptors. It constructs a candidate set of relevant lemmas: those dealing with corruption on earth, balance and moderation, obligations to fulfill covenants, justice between rich and poor nations, and generational fairness.

3. Reasoning graph construction: The tool then proposes a reasoning graph in which:

- Nodes represent lemmas (each node carries its formal condition, variables, and scriptural references).
- Edges represent logical or thematic links (for example, "lemma on fulfilling covenants" connects to "lemma on not entering unjust agreements").

4. Argument skeleton generation: From this graph, the tool derives possible argument skeletons, such as:

- "If we accept lemma A (obligation to avoid corruption) and lemma B (stewardship and preservation of balance), then policy C must reduce long-term harm to the environment below a specified threshold."
- "If we accept lemma D (fulfilling covenants) and lemma E (prohibition of unjust conditions), then we should support the agreement only if its financial terms do not create structural oppression."

These skeletons are not final opinions; they are proposed "paths of reasoning" that scholars can inspect, refine, accept, or reject.

Scholars in Control: Editing and Validating the Graph

The crucial feature of this tool is that it is assistive, not authoritative. Scholars can:

- Add or remove lemmas from the reasoning graph.
- Adjust the relationships between nodes, merging or splitting argument lines.
- Change assumptions about how certain variables (for example, harm indices or risk measures) are quantified in the real world.
- Attach weights to competing considerations (for instance, weighting preservation of life more heavily than economic convenience in a medical context).

As these adjustments are made, the tool updates the proposed argument skeletons and highlights where assumptions or lemma choices lead to different recommendations. Over time, councils can also save and annotate reasoning graphs, building a living archive of structured fatwā reasoning that future scholars can study and extend.

What This Example Shows

This example illustrates CQ as a cognitive exoskeleton for scholarly reasoning. The tool does not issue fatāwā, nor does it replace the need for deep knowledge of fiqh, usūl, and context. Instead, it helps scholars manage complexity by:

- Systematically retrieving relevant Qur'anic lemmas across the entire CQ corpus.
- Making implicit argument structures explicit, in the form of graphs and skeletons.

— Allowing scholars to see how accepting certain premises and lemmas logically constrains or broadens possible conclusions.

In an age where problems are increasingly multi-dimensional and data-rich, such a tool enables scholars to uphold the richness of the tradition while engaging with new realities in a structured, transparent, and reproducible way. CQ, in this setting, becomes the shared substrate on which human reasoning can be supported, clarified, and, when necessary, challenged and improved.

4. DATA, LEARNING ALGORITHMS, AND LEMMA-LEVEL ANALYTICS

The Computational Qur'an (CQ) is not only a theoretical modeling framework; it is also a data-driven research program. Once verses have been expressed as systems of variables, equations, and lemmas, they can interact with empirical data, learning algorithms, and user feedback loops. This allows an entirely new kind of dialogue between revelation and observation: scripture provides the structural form, while data helps estimate parameters, validate patterns, and highlight where models need refinement or extension.

In this section, we illustrate three classes of applications:

1. Parameter estimation from real-world data, where CQ models are calibrated against historical or contemporary patterns.
2. Learning and retrieving topic-specific lemma clusters across the entire CQ corpus for AI-assisted reasoning.
3. Using user interaction data to refine CQ itself, turning it into a living, expert-guided system that learns over time.

Each example emphasizes that CQ is not a closed, static catalogue of formulas. Instead, it is a structured, scriptural mathematical corpus that can be explored and improved using modern methods from statistics, machine learning, and graph analytics – always under the supervision of human scholars and subject-matter experts.

Example 7: Parameter Estimation from Real-World Data

Consider a research center that wants to test whether CQ-derived models of oppression and civilizational decline match historical data. The Qur'an repeatedly narrates the rise and fall of nations: unjust societies that persist in oppression and corruption eventually face collapse, while those that repent or reform can be granted respite. CQ encodes these patterns as dynamic models. The next step is to ask: do these models, when given reasonable parameters, line up with what we observe in real history?

CQ Ingredients

Suppose CQ contains a "civilizational decline" model derived from verses about past nations. Such a model might express the rate of decline or collapse as a function of an oppression index and reform dynamics, for example:

where:

- is a "collapse pressure" or "structural fragility" index at time t.
- is an oppression index: concentration of power, systemic injustice, violation of rights.
- is a repentance/reform index: genuine efforts to correct injustice, return to just norms, and repair societal harm.
- represents the sensitivity of collapse pressure to oppression (how strongly injustice accelerates decline).
- represents the effect of repentance/reform in reducing collapse pressure (how powerfully reform slows or reverses decline).

In the CQ corpus, themselves may be functions of other variables – wealth distribution, institutional justice, relief for the oppressed, and so on. But at a high level, this equation captures a core Qur'anic intuition: unchecked oppression pushes societies toward collapse, while sincere reform can defer or mitigate that trajectory.

Learning Pipeline: From Data to Parameter Estimates

To empirically test and calibrate such a model, the research center builds a learning pipeline with the following components:

1. Data collection: For each society or country in a historical or contemporary dataset, gather time-series indicators that can serve as proxies for:

— : measures of oppression, such as press freedom indices, corruption indices, human rights violations, political exclusion scores.
— : indicators of reform efforts, such as genuine anti-corruption campaigns, legal reforms, reconciliation processes, transitional justice mechanisms.
— Outcomes related to decline and collapse, such as incidence of civil conflict, institutional breakdown, state failure indices, or large-scale refugee outflows.

2. Model specification: Translate the CQ equation into a statistical or computational form suitable for estimation. For example, treat C(t) as a latent (unobserved) collapse pressure that influences observable outcomes, or approximate collapse events as points where C(t) crosses a threshold.

3. Parameter estimation: Use standard estimation methods – maximum likelihood, Bayesian inference, or gradient descent on a time-series loss function – to find values of β_1 and β_2 (and any additional parameters) that best fit the observed data. In practice, this may involve:

— Defining a loss function that penalizes discrepancies between predicted and observed collapse-related outcomes.
— Running optimization procedures to minimize this loss across the dataset.
— Incorporating prior beliefs (for instance, that β_1 should be positive and β_2 should be positive) in a Bayesian framework.

4. Goodness-of-fit and diagnostics: Evaluate how well the CQ model, with its estimated parameters, explains patterns across different societies and time periods. This includes checking residuals, predictive accuracy, and sensitivity to data choices.

Interpreting the Results

Once parameters are estimated, the research center can ask:

— Do societies with persistently high modeled indeed exhibit higher rates of collapse-related outcomes, as the CQ model suggests?
— Does the inclusion of (reform) significantly improve predictive performance, consistent with Qur'anic narratives that emphasize the power of repentance and reform?
— In which contexts does the model over-predict collapse, and where does it under-predict?
— Do these discrepancies point to missing variables (for example, external intervention, natural resources, or regional dynamics) that should be incorporated into extended models?

These questions do not "test" the Qur'an in the sense of validating or invalidating revelation. Rather, they examine whether our human interpretation and formalization of Qur'anic patterns – encoded in CQ – captures enough of the relevant dynamics to be empirically useful. Where the model aligns well with data, we gain confidence in its structure and parameter ranges. Where it misaligns, we learn that our modeling may bc incomplete or context-sensitive and needs refinement.

What This Example Shows

This example demonstrates that CQ models are not merely elegant theoretical constructs; they can be empirically calibrated. When a civilizational decline equation is fitted to real-world data, it becomes a bridge between scriptural structure and observed history. This creates a new type of scholarship:

- One that respects the Qur'an as the source of the model's form.
- One that uses data-driven methods to estimate how strongly different factors (oppression, reform, justice) operate in various contexts.
- One that treats misfit not as a challenge to revelation, but as an opportunity to improve our models and deepen our understanding of how Qur'anic principles manifest in complex historical processes.

In this way, CQ invites a disciplined, mathematically grounded dialogue between revelation and empirical reality – without reducing one to the other.

Example 8: Learning Topic Lemmas Across the Entire CQ

A second data-and-learning application focuses not on parameter estimation, but on organizing knowledge at scale. As CQ grows, it will contain thousands of lemmas: small modeling units with text, equations, variables, and tags. For an AI assistant – or even a human researcher – it becomes crucial to retrieve the right subset of lemmas for a given question. The challenge is to build systems that can quickly assemble the full CQ context on a theme, such as "oppression," "trust," "family obligations," or "stewardship of the environment."

CQ Ingredients

In CQ, every lemma includes:

- A textual explanation: the natural-language description of the concept and its Qur'anic basis.
- Mathematical content: equations, inequalities, dynamic systems, or logical forms.
- Variables: references to the UQVD (for Qur'an) and other dictionaries when applicable.
- Tags: themes, domains, verse references, and structural information.

Additionally, CQ maintains a graph structure linking lemmas to each other based on shared variables, shared themes, and shared verses. This lemma graph is a rich, high-dimensional object that can be explored using modern graph and embedding techniques.

Algorithmic Pipeline for Topic Lemma Retrieval

To build an AI that can pull together full CQ context on a topic, we construct the following pipeline:

1. Joint lemma embeddings: Train embedding models that represent each lemma as a vector in a joint space capturing:

- Theological content, via language models applied to the lemma's text.
- Mathematical structure, via encodings of its equations and variable patterns (for example, including features for the presence of derivatives, integrals, thresholds, or particular variable clusters).
- Topical tags, via one-hot or dense representations of themes and domains.

The result is an embedding where lemmas that "mean similar things" theologically and structurally are positioned near each other.

2. Query embedding and retrieval: When a user submits a query such as "oppression in employment contracts," the system:

- Encodes the query into the same embedding space, using its text and any structured metadata provided (for example, domain = commerce, subdomain = labor).
- Retrieves lemmas whose embeddings are closest to the query, with a bias toward those involving oppression variables (such as O_P, injustice indices) and contract-related variables (such as risk-sharing, wage fairness, exploitation metrics).

3. Graph-based ranking: To refine retrieval, we apply graph algorithms over the lemma graph. For example, a personalized PageRank centered on the most relevant lemmas can rank others by how central they are to the oppression-in-contracts theme – taking into account not only direct similarity, but also connectivity through shared variables and verses.

4. Building a lemma context package: The final output is a curated "lemma context package": a set of, say, 10–20 lemmas that together represent the CQ knowledge on the requested topic. Each lemma arrives with:

- Its equations and variables.
- Its textual explanation and Qur'anic references.
- Links to related lemmas in the graph.

This package becomes the context window for an AI model or a human analyst who wants to formulate an answer, draft a fatwā, or write a research note grounded in the full CQ treatment of that topic.

What This Example Shows

This example illustrates how CQ can be used to train and drive topic-specific AI reasoning. Unlike generic language models that rely on diffuse textual patterns, a CQ-based system:

- Operates over an explicitly structured, scriptural mathematical corpus.
  - Retrieves lemmas not only by textual similarity, but also by shared variables, equations, and themes.
- Presents a well-scoped, inspectable context package that scholars and users can review.

In practical terms, this means that when an ethical question arises – about oppression, trust, family obligations, or any other theme – an AI assistant can rapidly assemble the relevant CQ lemmas, equations, and explanatory notes. Human experts remain responsible for interpretation and judgment, but their work is amplified by the system's ability to surface the most pertinent parts of a very large corpus in a principled way.

Example 9: Learning from User Interactions (Closing the Loop)

A third learning dimension emerges when CQ is deployed on a platform such as ShareMax and used by a community of scholars, students, and practitioners. Every interaction – model building, variable linking, ontology validation, simulation adjustment – generates data about how people understand and refine CQ. Instead of ignoring this stream, we can treat it as a source of feedback for improving CQ itself, while keeping scholars firmly in control of canonical content.

CQ Ingredients

On a live platform, CQ-related logs might include:

- Which models users copy, adapt, or modify most often.
- Which parameters they tend to adjust, and in what directions.
- Results of ontology games like "Tinder-for-triples," where users accept or reject semantic triples involving variables, relations, and concepts.
- Outcomes of "Conflict Arena" debates, where alternative modeling choices or interpretations are compared and voted on.

All of this interaction data reflects how the community engages with CQ: where they find the models intuitive or confusing, where they push for refinements, and where emergent consensus forms.

Learning Pipeline: Reinforcement-Style Feedback

To tap this information, we can design a learning pipeline that treats certain interaction outcomes as implicit feedback signals:

1. Reward signals:

When many experienced users converge on a particular refinement of a model (for example, adding a missing variable or adjusting a functional form), we treat that as a positive reward for the new structure.

When ontology triples are consistently accepted by experts in games and discussions, they gain positive reinforcement; consistently rejected triples receive negative reinforcement.

2. Recommendation models:

Train models that, given the current CQ corpus and interaction history, can suggest likely useful new lemmas or variable linkages.

For example, the system might notice that many models dealing with a certain theme (say, trust in contracts) implicitly depend on a variable not yet formalized in the UQVD, and therefore recommend defining a new variable or lemma to capture it.

3. Human-in-the-loop curation:

Crucially, no automatic update to canonical CQ is made without scholarly review. The system presents candidate updates – new lemmas, revised equations, strengthened or weakened links – to a panel of scholars or editorial board.

Scholars then approve, modify, or reject these suggestions. Their decisions feed back into the learning system, improving its ability to distinguish between promising and misguided refinements.

4. Versioning and provenance:

- Every accepted change is versioned and recorded with provenance, indicating which interactions and expert decisions led to it. This ensures transparency and traceability of CQ's evolution over time.

What this example shows

This final example presents CQ as a living system that learns from its community. Algorithms observe patterns in how users interact with models and ontologies, then propose improvements. Experts vet these suggestions, ensuring that the evolution of CQ remains anchored in sound scholarship and responsible interpretation.

The result is a virtuous cycle:

—— CQ provides structured models and lemmas.
—— Users apply, critique, and extend those structures in real tasks.
—— Their interactions generate feedback signals.
—— Learning algorithms distill those signals into candidate refinements.
—— Scholars curate the system's growth, preserving integrity while benefiting from community insight.

In this way, the Computational Qur'an is not frozen at the moment of its first publication. It continues to mature – mathematically, conceptually, and practically – as it is used. The combination of data, learning algorithms, and lemma-level analytics turns CQ into a dynamic, communal project: deeply rooted in revelation, rigorously formal in its methods, and responsive to the needs and insights of each new generation of users.

FALSIFIABILITY, REPRODUCIBILITY, AND MODEL FAILURE IN THE COMPUTATIONAL QURAN

A central claim of the Computational Quran is that scriptural interpretation can be made more explicit, structured, and testable. This does not mean that divine revelation is reduced to mathematics, nor that human scholarship is replaced by computation. Rather, it means that interpretive claims can be expressed in forms that are clear enough to examine. A model should show its variables, assumptions, relationships, thresholds, and expected outcomes. Once these are made explicit, the model can be tested. It can be challenged. It can even fail.

This is where the scientific concepts of **falsifiability** and **reproducibility** become important. Falsifiability means that a model must be capable of being shown wrong, incomplete, or inadequate under specific conditions. If a model can explain every possible result after the fact, then it is not truly testable. It may be rhetorically useful, but it is not scientifically disciplined. Reproducibility means that another scholar, using the same text, variables, assumptions, and procedure, should be able to repeat the analysis and see whether the model succeeds or fails in the same way. Together, falsifiability and reproducibility help move interpretation from private assertion toward transparent method.

A simple example can be drawn from the Quranic concept of repentance. Suppose an initial Computational Quran model defines repentance as the combination of three variables:

[Repentance = Recognition + Regret + BehavioralCorrection]

At first glance, this appears reasonable. A person recognizes wrongdoing, feels regret, and changes behavior. This model may work for many ordinary cases where a person commits a wrong, becomes aware of it, turns back to God, and reforms his or her conduct. It captures repentance as a moral transition from error to correction.

But the model must be stress-tested. A strong test case is Pharaoh's late declaration of belief at the moment of drowning. In the Quranic account, Pharaoh resists divine signs, rejects prophetic warning, oppresses the Children of Israel, and pursues them until the sea closes upon him. Only when drowning becomes unavoidable does he declare belief. If the initial model is applied mechanically, it may classify Pharaoh's final statement as valid repentance. The model detects recognition: Pharaoh now acknowledges the truth. It may infer regret from the crisis. It may even interpret the verbal confession as a corrective turn.

This is a failed simulation.

The failure is not trivial. It reveals that the original model is missing essential variables. Repentance is not merely recognition at any moment. It must be connected to moral agency, sincerity, timing, and the possibility of transformation. Pharaoh's statement occurs after the consequences of his actions have become inescapable. It comes when the window for meaningful behavioral correction has closed. Therefore, the model wrongly treats late recognition under compulsion as equivalent to sincere repentance under conditions of agency.

The model must now be revised:

[RepentanceValidity = f(Recognition, Regret, BehavioralCorrection, Timing, Sincerity, PreDeathOpportunity)]

This revised model is more faithful to the Quranic structure. It does not discard recognition, regret, or behavioral correction, but it adds the missing conditions that determine whether repentance is valid. Timing matters because repentance after accountability becomes visible is different from repentance while moral choice remains open. Sincerity matters because verbal confession under duress may not represent inward transformation. PreDeathOpportunity matters because repentance requires the possibility of changed life, not merely final acknowledgment at the edge of punishment.

This example illustrates why model failure is valuable. The failed simulation does not weaken the Computational Quran. It strengthens it. It shows that CQ models are not decorative equations placed beside verses for intellectual effect. They are working interpretive structures. They can be tested against difficult cases. They can be found insufficient. They can be corrected.

The same test can also be reproduced. Another scholar can take the initial repentance model, apply it to ordinary cases of repentance, then apply it to the Pharaoh drowning episode, observe the misclassification, and evaluate whether the revised model resolves the problem. This makes the interpretive process more transparent. Instead of simply arguing rhetorically that Pharaoh's repentance was invalid, the model shows which structural conditions are missing and why they matter.

In this way, falsifiability and reproducibility become essential to Computational Quranic analysis. They allow models to mature through stress-testing. They expose hidden assumptions. They discipline interpretation. Most importantly, they demonstrate that computational modeling is not a replacement for tafsir, theology, or moral reasoning. It is a method for making interpretive reasoning visible, testable, and improvable.

Chapter 10
Applied Quran:
Reformulating Social Sciences

ABSTRACT

This chapter argues that the Quran can function not only as a sacred text for devotion but as a foundational framework for social science – shifting the work from interpretation to application. It proposes a four-layer structure for Quran-based social science consisting of foundational axioms (such as fitrah, justice, and balance), social objectives, testable causal propositions, and empirical simulation, with the Computational Quran serving as the bridge that translates revelation into variables, relationships, and measurable indicators a cognitive machine can reason over. The authors maintain that all social science begins with assumptions, and that a Quranic framework simply makes its assumptions explicit and tests what follows. The chapter then demonstrates the method through two worked examples built on open-source intelligence and Computational Quran cluster models: a cross-national analysis of institutional hypocrisy and oppression, and a systems-level analysis of social cohesion, deception entropy, and societal decline.

INTRODUCTION

In this chapter we argue for something much bigger than theology alone. We say the Quran can now be studied as more than a sacred text as it can also serve as a framework for studying society, human behavior, institutions, and social outcomes in a disciplined way.

The key shift is from interpretation to application. Traditional theology asks what God commands and how verses should be understood. This new approach asks what the Quran teaches about human nature, justice, corruption, trust, decline, resilience, and social flourishing – and how those ideas can be modeled, tested, and applied.

DOI: 10.4018/979-8-3373-6940-2.ch010

That is what Quran-based social science means. It treats revelation not only as an object of belief, but also as a source of assumptions, causal ideas, goals, and standards for analyzing real societies. Computational Quran (CQ) is necessary because sacred text alone is not enough for machine-enabled science. To build a scientific system, Quranic guidance must be translated into variables, relationships, state changes, causal pathways, measurable indicators, and testable propositions.

CQ IS THE BRIDGE

CQ is the bridge between revelation and scientific machine reasoning. It allows a cognitive machine not just to quote verses, but to reason, model, compare, simulate, test, recommend, and discover within a Quranic framework. We propose four main layers for Quranic social science.

⇒ First are foundational axioms, such as fitrah, moral accountability, justice, balance, and the link between moral disorder and social disorder. These serve as first principles, just as other social theories begin with autonomy, class conflict, or utility.

⇒ Second are social objectives, or the goals of a healthy society. Instead of measuring success only by growth, efficiency, or consumption, a Quranic framework would evaluate whether society protects and strengthens life, intellect, family, property, and faith or moral meaning.

⇒ Third are causal propositions. The Quran does not only teach values; it also suggests relationships such as injustice causing instability, arrogance blinding judgment, corruption spreading through abuse of power, and truthfulness strengthening trust. These can be turned into hypotheses and tested.

⇒ Fourth is empirical testing and simulation. A CQ-governed machine could use real-world data, compare communities across time, track social variables, identify threshold effects, test Quranically predicted patterns, and simulate policy consequences while keeping the Quranic moral core intact.

A machine becomes truly "CQ-governed" only when Quranic structure shapes its reasoning. This means its concepts include categories like justice, oppression, trust, corruption, gratitude, hypocrisy, stewardship, and repentance. It also means the machine uses Quranic causal logic, Quranic standards of success, Quranic constraints, and interpretable reasoning tied to explicit modeled structures.

APPLICATIONS IN REAL LIFE

Let's see how this could work in real fields:

⇒ In poverty and inequality, the machine could study not just low income, but wealth concentration, blocked circulation, household stress, trust, institutional legitimacy, and social fracture.

⇒ In family and social fabric, it could examine marriage stability, child development, intergenerational support, loneliness, elder care, and demographic weakness. The Quranic idea is that when kinship and family responsibility decline, society suffers long-term hidden damage.

⇒ In governance and legitimacy, it could go beyond efficiency and state capacity. It could measure justice in adjudication, corruption, abuse of entrusted power, elite arrogance, truthfulness in authority, and protection of the weak. A society may look functional on paper while decaying morally underneath.

⇒ In mental health and existential stability, it could expand the model of the human being. Instead of reducing distress only to biology or environment, it could include meaning, remembrance, gratitude, despair, heedlessness, repentance, and moral coherence.

⇒ In crime, deviance, and corruption, it could model not just incentives and enforcement, but also greed, arrogance, moral desensitization, permissiveness, predatory elites, and spiritual hardening. This gives a richer explanation of how corruption spreads and when injustice becomes normalized.

⇒ In development economics, it would ask a different question from modern development theory. Instead of only asking whether GDP and investment are rising, it would ask whether society is becoming more just, balanced, stable, dignified, and morally coherent.

⇒ In social cohesion and civilizational resilience, it could treat justice as a stabilizer and oppression as a fracture point. It could model how hypocrisy, arrogance, covenant-breaking, and moral decay weaken a society before material collapse becomes obvious.

We think of this as a scientific workflow, not just a belief claim. First, extract a Quranic principle. Second, formalize it computationally. Third, derive hypotheses. Fourth, test them with data. Fifth, simulate interventions. Sixth, evaluate results using Quranic goals.

The response to critics is that all social science already begins with assumptions. Rational choice, liberal theory, Marxism, and functionalism all start with hidden or open assumptions about human beings and society.

A Quranic framework simply makes its assumptions explicit and then tests what follows from them.

The cognitive machine is essential because it makes this project scalable. A human scholar can do part of the work, but a CQ-governed machine could map thousands of verses, connect concepts across domains, generate hypotheses, test them against data, compare societies, run simulations, and produce interpretable outputs.

The deepest claim of the chapter is that CQ allows revelation to enter the field of social science in operational form. In this model, the Quran becomes not only a source of morality and devotion, but also a source of social theory, behavioral theory, governance theory, development theory, and civilizational diagnosis.

The central idea is this: a CQ-governed cognitive machine could make possible a new Quran-based social science that studies society through Quranic concepts, tests those ideas with data, and evaluates success by whether human life is protected, elevated, and kept in moral balance.

THE APPLIED QURAN – NOT JUST THEORY, BUT GLOBAL TRANSFORMATION WITH QURAN

Quran describes its own purpose. It presents itself as guidance, healing, mercy, a criterion for truth and falsehood, and a reminder that awakens what is already present in the human soul.

The Quran says it is meant to transform people, not just inform them. It is described as guidance for the mind and soul, a force that brings people from darkness into light, a healing for what is in the heart, and a standard for judging right from wrong.

It also says that reflection is necessary. The Quran may be clear, but its deeper purpose is realized through *tadabbur*, meaning serious reflection, understanding, and thought.

Let us also understand what social science tries to do. Social sciences study human behavior, relationships, institutions, power, culture, psychology, and society in a systematic way, much like hard sciences study the physical world.

Their main goals are fourfold.

1. They try to describe what is happening in society, explain why it is happening, predict what may happen next, and improve human well-being through better understanding and policy.
2. Social sciences also challenge assumptions. They help us see that many things we take as natural may actually be shaped by history, culture, institutions, and environment.
3. We need social sciences because human beings are complex. Unlike chemical reactions, people are influenced by emotions, beliefs, culture, and free will, so society cannot be understood through simple observation alone.

4. Social sciences often move between assumption and discovery. Sometimes they begin with a model they take for granted, such as democracy or rational self-interest. At other times, they test those assumptions and discover their limits.

This means social science is never fully neutral. It often contains a hidden goal, or *telos*, such as liberty, equality, efficiency, order, or stability. Researchers then build models around those goals.

The proposed shift is to build a Quranic social science. Instead of using only secular frameworks such as liberalism, Marxism, or rational choice theory, this approach uses the Quran as a foundational theory for understanding society.

In this model, Quranic concepts become the starting axioms. Human beings are understood through ideas such as *fitrah* (innate moral nature), *mithaq* (moral covenant), *adl* (justice), and *mizan* (balance). These function like first principles in the analysis of society.

The goals of society are also redefined through a Quranic lens. Instead of measuring success only by GDP, growth, or efficiency, success is judged by whether society protects life, intellect, family, property, and faith or meaning.

This is where the scientific side begins. Quranic ideas are treated as hypotheses that can be tested. For example, if the Quran suggests that concentration of wealth harms society, that claim can be examined using inequality data, trust data, fragility data, and historical outcomes.

This creates a real methodological framework. Revelation provides the theory, and observation plus data provide the testing process. In that sense, the Quran becomes a theoretical framework for social science, not only a text for exegesis.

A major example is the idea of social balance. The Quranic concept of *mizan* can be translated into measurable indicators such as wealth inequality, access to justice, and institutional balance. Researchers can then test whether extreme imbalance leads to unrest or social breakdown.

Another example is the link between inner change and outer change. The verse that says God does not change the condition of a people until they change what is in themselves can be read as a social law connecting personal and collective transformation to broader societal outcomes.

We introduce a "Quranic Loop" for doing research. First identify a Quranic principle, then define variables, analyze data, compare social systems, and propose reforms that better align society with Quranic goals.

We provide a comparison with conventional social science. Conventional approaches usually begin with human observation alone, while Quranic social science begins with revelation plus observation.

Conventional models may treat humans mainly as biological or economic actors, while the Quranic model treats them as spiritual and moral beings with material needs.

Several practical examples are available. One is economic life, where Quranic ideas such as opposition to exploitation and the circulation of wealth can be tested against real systems. Another is social cohesion, where the Quranic idea of *ta'aruf* – meaningful mutual recognition between groups – can be used to study polarization, segregation, and echo chambers.

In that example, diversity is treated as socially functional, not accidental. The Quranic idea is that peoples and tribes exist so they may know one another, not isolate from one another. This becomes a testable hypothesis about whether societies with stronger cross-group ties are healthier and more resilient.

Quranic concepts can be mapped onto social science language. For example, *fitrah* can function like a theory of human nature, *'adl* like systemic justice, *ummah* like social capital or collective identity, and *tadabbur* like disciplined reflection and analysis.

The deeper claim is that the Quran can be used as a predictive and explanatory map for society. Rather than treating it only as a holy text for devotion, this framework treats it as a source code for understanding social order, human behavior, institutional health, and civilizational success.

In simple terms, we argue for a major shift: move from only explaining what the Quran means, to using the Quran's logic to build models, test hypotheses, and design better societies.

The overall conclusion is that Quranic social science is possible. It would combine revelation, structured concepts, empirical testing, and practical reform, allowing the Quran to function not only as guidance for personal faith, but also as a framework for understanding and improving human society: The Applied Quran.

APPLIED EXAMPLES OF COMPUTATIONAL QURAN

The preceding chapters have laid out the epistemological argument for why Quranic axioms constitute a valid and rigorous foundation for social science. This chapter puts those tools to work. Rather than continue at the level of theory, we walk through a complete applied example – from a specific Quranic passage, to a formal model, to a dataset, to a result – to show what Quranic Social Science actually looks like in practice.

Example 1: Oppression Across Nations

Imagine you are an analyst tasked with evaluating the institutional integrity of a government – any government. You want to know whether it is actually doing what it claims to do. You search for a measurement framework and find plenty of options:

corruption perception indices, rule-of-law scores, governance effectiveness ratings. You pick one, run your numbers, and produce a result. But here is the problem you may not have fully confronted yet: every one of those frameworks was designed by someone, and that someone had a theory – often unstated – of what institutional failure looks like, what counts as harm, and whose conduct is worth measuring carefully. If your framework was built by and for Western liberal institutions, it will find what those institutions defined as the problem. It will measure bribery in procurement offices. It will not measure how many civilians died from aerial strikes authorized at the highest levels of a constitutional democracy. It will not measure the monetary sovereignty extracted from former colonies through currency arrangements still in force today. It will not ask whether a government's elaborate stated commitment to international law has any relationship to its actual behavior. These are not measurement gaps caused by missing data. They are structural choices about what matters – and if you accept those choices uncritically, you have already accepted the conclusions before you begin the analysis. What follows in this chapter shows a different way to proceed: starting not from what existing institutions have found convenient to measure, but from what the Quran itself identifies as the structural sources of harm to human civilization, and building measurement categories from those foundations upward. You will see that when you do this – when you build your model from primary-source open-source intelligence (OSINT) rather than packaged indices, and when you deploy multiple Quranic harm models rather than a single composite score – the results are analytically more honest, more discriminating, and more useful to any analyst who actually wants to know what is happening rather than what a preferred framework was designed to reveal.

The model we apply here is drawn from Cluster 2N of the Computational Quran source text (pages 45–46), which formalizes a passage from Surah Al-Baqarah. In that passage, the Quran presents an archetype: a figure whose speech is compelling, whose stated commitments are pious, but whose actions generate corruption. The gap between what is claimed and what is done is identified not merely as dishonesty but as a structural source of *fasad* – the compounding deterioration of social trust, stability, and moral coherence. The Computational Quran translates this archetype into two functions (Naqvi & Naqvi, 2026).

$$D(t) = \| S(t) - A(t) \|$$

$$EC(T) = \int_0 \varphi \cdot 1\{ A(t) \in \text{harmful acts} \}\, dt$$

The first function, D(t), measures the divergence at any given time between what an institution states – its pledges, its declared commitments, its formal self-presentation – and what it actually does, as measured through independent evidence.

The second, EC(T), accumulates the harm generated by that divergence over time. Together they capture what the Quranic text describes: that the damage of institutional hypocrisy is not a single event but a trajectory, and that trajectories accumulate.

To apply these models, we need two operationalized measures: S(t) and A(t). For this illustration we construct S(t) from each institution's formal stated commitments – treaty ratifications, official strategy documents, declared mandates, and public governance pledges – scored on a 0–100 scale where 100 represents maximum stated ambition. We construct A(t) not from pre-packaged institutional indices but from primary-source evidence: conflict documentation, investigative research records, parliamentary investigation outcomes, labor exploitation data, and information environment assessments. The distinction matters. Existing governance indices measure what their designers chose to measure. Our A(t) measures what the Quranic framework's own harm categories require us to measure: *zulm* (direct harm to persons), structural exploitation of the vulnerable, concealment of truth, and accountability for wrongdoing.

We apply this to a set of seven illustrative country profiles – labeled Country A through Country G – drawn from different regions, political systems, economic scales, and cultural contexts. The profiles are composites designed to represent the full range of governance contexts a real deployment of this framework would encounter. No real country is named here; the point is methodological demonstration, not political adjudication.

Table 1. Illustrative country profiles and domain scores (0–100, higher = greater harm)

Country	D1: State Violence	D2: Exploitation	D3: Deception	D4: Hypocrisy D(t)	D5: Accountability	Composite
Country A	35	28	32	56	38	37.7
Country B	22	52	58	44	52	44.1
Country C	66	62	72	68	63	66.3
Country D	66	72	62	72	62	67.0
Country E	72	58	68	78	68	69.0
Country F	62	83	74	58	72	69.3
Country G	82	78	78	52	78	73.8

Each domain in the table corresponds to a distinct model in the Computational Quran framework. D1 (State Violence) draws on the Cluster 11A model of ethical decay from harm to the vulnerable and the Cluster 4B governance legitimacy metric. D2 (Structural Exploitation) applies the Cluster 2W *ribā* criterion: economic arrangements that guarantee benefit to the powerful party while imposing all cost and risk on the weaker constitute the same structural pattern as forbidden extraction, regardless of what they are called. D3 (Deception Entropy) draws on the Cluster 3D and 3J models, measuring the disorder introduced into a society's epistemic environment by information suppression, historical denial, and coordinated disinformation. D4 is the Cluster 2N divergence function itself. D5 measures the accountability deficit modeled in Clusters 4B and 3D: the degree to which documented harms generate prosecutions, reparations, and structural reform.

The results illustrate several properties of the framework that are worth drawing out explicitly.

⇒ First, Country E – a high-income constitutional democracy – scores the highest D4 hypocrisy score in the sample at 78. This is not a paradox. It reflects the model's structural logic: an institution that declares itself the global guardian of human rights and the rules-based international order generates an extremely high S(t), and when its behavioral record includes documented civilian deaths from aerial operations, mass incarceration, and systematic surveillance of allied governments, the resulting divergence is proportionally large. The Quranic model does not award credit for eloquence. It measures the gap.

⇒ Second, Country A – which has the lowest composite score in the sample – nevertheless carries a D4 divergence score of 56, meaningfully above its D1 direct violence score of 35. This points to a specific structural finding: its greatest failure is not in what it directly does but in the gap between its extraordinarily emphatic stated commitments to international law – built into its foundational national identity – and documented episodes of enabling mass civilian casualties through supply chains and selective cooperation with international accountability mechanisms. Low overall harm and high stated values can still generate significant divergence.

⇒ Third, the table reveals a cluster at the composite level: Countries D, E, and F all score within two points of each other (67.0, 69.0, 69.3) despite representing a Western liberal democracy, a major constitutional republic, and a Gulf state respectively. Standard governance indices would place these three countries in entirely different categories. The Quranic framework finds them structurally comparable because they share the same combination of documented direct harm, exploitation of dependent populations, and accountability deficits – even though the specific mechanisms differ across all three.

⇒ Fourth, note that Country G scores the highest composite (73.8) despite having the lowest D4 hypocrisy score (52). This reflects that the Quran's condemnation of oppression and exploitation does not depend on the oppressor being a hypocrite. Country G claims relatively little; it diverges relatively little from what it claims; but the absolute scale of documented harm it generates – particularly in domains D1 and D5 – produces the highest overall score in the sample. The framework correctly identifies that hypocrisy is one mechanism of institutional failure, not the only one.

What this example demonstrates is that Quranic Social Science produces analytically discriminating results across a genuinely diverse global sample. It rewards Country A's relatively strong institutional performance. It finds the structural equivalence between Countries D, E, and F that convenience would obscure. It identifies Country G's absolute harm independently of its modest stated ambitions. It does all of this using measurement categories derived from the Quran's own descriptions of how institutions fail – not from the categories that powerful institutions have found convenient to make measurable.

This is the operational meaning of the claim that the Computational Quran provides a foundation for a new social science. It is not that the Quran replaces data with scripture. It is that the Quran provides the harm taxonomy – what counts as injustice, what constitutes exploitation, what makes institutional speech meaningful or hollow – and that taxonomy, formalized mathematically and grounded in OSINT evidence, produces analyses that existing frameworks structurally cannot. A framework built on Western governance indices will find what Western governance institutions have chosen to measure. A framework built on Quranic categories will find what the Quran identifies as structurally destructive to human civilization – wherever that destruction is occurring, and whoever is responsible for it.

The exercise in this chapter is an illustration. In a fully deployed Quranic Institutional Integrity Engine, these five domain scores would be computed continuously from live OSINT feeds, updated as new evidence emerges, and integrated across all 114 surahs' worth of models rather than the five clusters used here. The chapter that follows begins that broader integration, connecting the hypocrisy model developed here to the memory-decay and social-cohesion models developed in Chapter 2 to show how institutional divergence propagates into the social fabric over time.

Chapter Note on Data

The country profiles in Table 1 are illustrative composites used to demonstrate the framework's analytical properties. The domain scores are synthetic, constructed to represent the range of governance contexts the framework would encounter in a real deployment. The OSINT methodology used to construct A(t) in a live analysis – drawing on conflict databases, UN special procedure documentation, labor exploitation research, digital surveillance forensics, and investigative journalism archives – is described in detail in the accompanying technical appendix.

Models Applied in This Chapter

Cluster 3J – *Social Equilibrium Enforcement, Fragmentation Resistance, and Deception Field Detection* | Surah Āl-Imrān 3:102–120 | pp. 88–90

Cluster 3I – *Action-Based Truth Validation, Information Trust Degradation, and Cohesion Potential Fields* | Surah Āl-Imrān 3:92–101 | pp. 68–69

Cluster 2A – *Revelation and Belief Archetypes* | Surah Al-Baqarah 2:1–20 | pp. 7–8

Example 2: Social Decline

The previous example asked you to evaluate what an institution claims against what it actually does. That is a powerful diagnostic, but it operates on a single actor in relative isolation. Now consider a harder problem: you are not evaluating one government or one institution. You are trying to understand what is happening to an entire society. The question is no longer whether a specific actor is lying – it is whether the social environment itself has become so saturated with manipulation, division, and competing narratives that the community can no longer function as a coherent reasoning unit. You will recognize this condition immediately. You have seen it in post-election environments where coordinated disinformation campaigns have made it functionally impossible for citizens to agree on basic facts. You have seen it in conflict zones where information operations run by external actors are deliberately accelerating ethnic and sectarian tensions that already existed. You have seen it on social media platforms where algorithmic amplification rewards outrage over accuracy until the population that was once persuadable has fragmented into irreconcilable camps. What you probably have not had is a rigorous, computable framework for measuring when a society has crossed the threshold from stressed but recoverable into structurally vulnerable – and for identifying which variable is the actual driver, because the intervention you design depends entirely on that answer. The models in this chapter give you that framework. They draw on three clusters from the Computational Quran that together describe how social cohesion degrades, how external manipulation accelerates that degradation through measurable entropy in the information environment, and – most counterintuitively – why the most dangerous condition is not when people lack access to true information, but when truth is accessible and community unity has collapsed anyway, leaving accurate knowledge with no functional social architecture through which it can produce coordinated action.

The previous worked example applied Cluster 2N's hypocrisy metric to institutional governance – measuring the gap between what states declare and what they do. That model operates at the level of individual and institutional actors. This chapter applies a different set of models that operate at the level of social systems:

how communities hold together under pressure, how that cohesion degrades when external manipulation enters the information environment, and at what point the degradation crosses a threshold beyond which a society becomes structurally vulnerable to collapse from within.

The source material is Surah Āl-Imrān, verses 3:102–120, formalized as Cluster 3J of the Computational Quran (pages 88–90). The passage addresses a community facing simultaneous internal fractures and external bad actors who are actively working to deepen those fractures. The Quran's structural analysis of this situation is precise: it identifies unity as a field with a gravitational attractor, fragmentation as a force acting against that field, and external manipulation as a source of entropy – disorder – in the community's epistemic environment. When that entropy exceeds a threshold, the community loses the ability to distinguish reliable from unreliable information, and its vulnerability to further manipulation becomes self-reinforcing.

Three models from the Computational Quran framework work together to formalize this dynamic. The first, from Cluster 3J, defines social cohesion as a function of the unity attractor and collective behavior, offset by fragmentation risk:

$$SC(t) = f(UA, CB) - FR$$

$$dFR/dt = \alpha \cdot \|SD\| - \beta \cdot \|UA\|$$

SC(t) is social cohesion at time t. UA is the unity attractor – the shared normative framework, values, and commitments that hold a community together. CB is collective behavior, the degree to which the community is actually acting in accordance with its stated values. FR is fragmentation risk, which grows when sectarian diffusion SD is strong and shrinks when the unity attractor is reinforced. When SC(t) approaches zero, the Cluster 3J model identifies a critical threshold: the system becomes susceptible to sectarian diffusion – the propagation of divisive narratives through social networks.

The second model, also from Cluster 3J, formalizes the role of external manipulation in accelerating this process through a deception entropy function borrowed from information theory:

$$DE = H(EM) = -\Sigma\ p\ \log p$$

$$DE > \theta \Longrightarrow \text{Trigger Cognitive Defense Protocols}$$

DE is deception entropy. EM is the external manipulation vector – the intensity, variety, and coordination of disinformation entering the community's information environment. H(EM) measures the disorder that manipulation introduces: how scrambled the probability distribution of true and false signals has become. A community receiving a clean, internally consistent information environment has low entropy. A community being bombarded with coordinated disinformation, algorithmically amplified divisive content, and coordinated inauthentic behavior faces high entropy – it cannot efficiently separate signal from noise. When DE exceeds the threshold θ, the model activates cognitive defense protocols: individuals and institutions must expend resources distinguishing truth from manipulation rather than acting on it.

The third model draws from Cluster 3I (pages 68–69), which formalizes the counterintuitive condition the Quran identifies as the most dangerous: the state in which truth is known but community unity has collapsed anyway.

$$TN(t) = T_0 - \int DC(\tau)\, d\tau \text{ [Information Trust Degradation]}$$

$$TK = 1 \wedge \text{Disunity} \uparrow \Longrightarrow \nabla V(t) = \text{Vulnerability Field}$$

The trust degradation function TN(t) measures how network trust erodes as data corruption DC accumulates over time – each fabricated narrative, each manipulated data point, each false attribution reduces the baseline trust from which a community reasons. The vulnerability gradient is the more striking result: it states that when truth is known (TK = 1, meaning correct information is accessible) but disunity is rising anyway, a vulnerability field emerges across the community. This is the post-truth condition formalized: the problem is no longer that people lack access to accurate information. It is that the social architecture required to act on that information coherently has been degraded.

To understand why a third model is needed to complete the picture, we draw on Cluster 2A's belief archetype framework (pages 7–8). Cluster 2A models three distinct human response patterns to a shared information environment: the believer who receives and acts on guidance coherently, the outright rejector whose receptivity has collapsed, and the hypocrite whose response oscillates without converging:

$$d^2H(t)/dt^2 + \alpha\ dH(t)/dt + \beta H(t) = SQ(t) - DT(t)$$

$$dHN/dt \sim \sin(\omega t) \text{ [non-convergent oscillation under manipulation]}$$

Under high deception entropy, a community's population distribution across these three archetypes shifts. The coherent responders remain anchored by strong internal frameworks. The outright rejectors become noise in any collective decision. But the oscillating archetype – modeled as a sinusoidal, non-converging response – is the vulnerable population: people who are neither reliably aligned nor permanently opposed but who respond unpredictably to the manipulation signal. As deception entropy DE rises, this oscillating population grows, and with it the community's collective inability to respond coherently to any shared challenge.

Table 2. Illustrative social cohesion and deception entropy profiles – eight platform environments

Environment	SC(t)	DE Score	FR Rate	TN(t)	Oscillating Pop. %	Status
Platform A (High-income democracy, social media dominant)	68	31	Low	74	18%	Stable
Platform B (Post-election period, coordinated campaigns)	52	54	Rising	61	29%	Warning
Platform C (Active conflict zone, state-linked IO)	34	72	High	42	41%	Critical
Platform D (Authoritarian state, domestic censorship)	41	48	Medium	38	22%	Warning
Platform E (Fragile democracy, ethnic polarization)	29	79	High	31	47%	Critical
Platform F (Stable high-trust society, low IO exposure)	83	18	Low	88	9%	Stable
Platform G (Gulf state, algorithmic amplification)	44	61	Rising	47	33%	Warning
Platform H (Post-coup society, competing narratives)	21	88	Critical	24	58%	Collapse Risk

Synthetic indicators constructed from OSINT methodology. SC: social cohesion index (100 = maximum unity). DE: deception entropy score (100 = maximum disorder). FR: fragmentation risk rate. TN: network trust level. Oscillating Population: estimated share of population in non-convergent belief response.

The table presents eight synthetic information environment profiles constructed to illustrate how the three models interact across different conditions. The profiles draw on OSINT methodology patterns observed in contexts involving coordinated inauthentic behavior campaigns, algorithmic amplification of divisive content, and state-linked information operations – without specifying real platforms or countries by name.

Several dynamics visible in the table are worth drawing out. Platform F represents the stable baseline: high SC(t), low DE, high TN(t), and only 9% of the population in the oscillating archetype. This is the condition the Cluster 3J models identify as resilient – an active unity attractor, strong collective behavior, and a low-entropy information environment. Platform H represents the opposite extreme: a DE score of 88, an oscillating population of 58%, and social cohesion so degraded (SC=21) that the model classifies it as facing collapse risk. The trust network (TN=24) has been so eroded by accumulated data corruption that even correct information cannot function effectively.

The most analytically instructive case is the transition from Platform B to Platform C. Between these two environments, DE rises from 54 to 72, SC drops from 52 to 34, and the oscillating population grows from 29% to 41%. This trajectory – not a single event but a compounding process – is exactly what the integral form of the trust degradation model captures. Network trust TN(t) does not collapse from a single disinformation campaign; it degrades continuously as each fabricated narrative integrates into the cumulative corruption stock. The Cluster 3I formula $TN(t) = T_0 - \int DC(\tau)d\tau$ formalizes why societies that appear resilient in any single snapshot can reach collapse conditions through gradual accumulation.

The post-truth vulnerability gradient – $TK = 1 \wedge \text{Disunity} \uparrow \implies \nabla V(t) = \text{Vulnerability Field}$ – is the Quran's most counter-intuitive structural insight in this domain. It is not ignorance that creates maximum vulnerability. It is the combination of accessible truth and fractured social architecture that produces the vulnerability field. A community in which accurate information is available but trust has been destroyed cannot coordinate on that information. This is the condition that information operations are specifically designed to produce.

The practical implication for AI system design is direct. A Quranic Institutional Integrity Engine drawing on these models would not simply flag high-DE environments as 'disinformation problems' requiring content removal. It would compute SC(t) and TN(t) simultaneously with DE, identifying whether the primary intervention target is the information environment (high DE, stable SC), the trust network (declining TN with moderate DE), or the oscillating population distribution (growing sinusoidal responders). Each diagnosis maps to a different class of intervention, and the Cluster 3J model's threshold condition $DE > \theta$ provides a computable trigger for when passive monitoring must give way to active cognitive defense.

Two further connections to the broader CQ framework are worth noting. First, the social cohesion model here directly extends the hypocrisy analysis from the previous chapter. Cluster 2N's speech-action divergence D(t) was applied to institutions; that same divergence, when it becomes the dominant signal in a society's information environment, is precisely the external manipulation vector EM that drives deception entropy in the Cluster 3J model. An institution whose speech-action divergence is high is, through that gap alone, contributing entropy to the broader epistemic field of any community that believes it. Second, the oscillating population archetype from Cluster 2A – the nifāq pattern of non-convergent response – is not merely a theological observation. It is a behavioral profile with measurable correlates: high susceptibility to framing effects, low epistemic stability across different information sources, and disproportionate amplification of divisive content in social network dynamics. These are precisely the populations that information operations target.

What the Computational Quran offers here that no conventional disinformation framework currently provides is a unified causal model: social cohesion, trust degradation, deception entropy, and population-level belief archetype distribution are not separate phenomena requiring separate analytical tools. They are linked by a single structural account of how epistemic environments and social bonds co-determine each other. The model does not merely describe the problem. It computes it.

Chapter Note on Data and Citation

The platform profiles in Table 2 are synthetic composites constructed to illustrate the framework's analytical range. They draw on OSINT methodology patterns including the EU DisinfoLab categorization framework for coordinated inauthentic behavior, the Oxford Internet Institute's Computational Propaganda Project typology, and Access Now's documented internet shutdown and information operation patterns. No real platform or country is directly identified. Full citation of source models: *Cluster 3J (pp. 88–90), Cluster 3I (pp. 68–69), Cluster 2A (pp. 7–8).*

REFERENCES

Naqvi, A., & Naqvi, N. (2026). *The Computational Quran for the AI Era: Selected Mathematical Models of Divine Guidance*. American Institute for AI and Quantum.

Section 4
Applications, Implications, and Future Directions

Chapter 11
From Scripture to Machine:
Integrating the Computational Quran With AI

ABSTRACT

Chapter 11 explains how the Computational Quran becomes operational within artificial intelligence by translating revelation into variables, equations, graphs, and dynamic systems that machines can reason over. It shows how CRS frameworks – UQVO variables, verse-level models, and thematic graphs – form a Quranic world model that AI can query, simulate, and constrain its decisions with. The chapter presents integration patterns for LLMs, neuro-symbolic systems, ethical engines, spiritual advisors, and societal simulations, emphasizing transparency, governance, and theological accountability. It argues that modeling scripture mathematically enables AI to understand moral and spiritual dynamics rather than merely quoting text, positioning revelation as an active computational architecture guiding human–machine interaction.

INTRODUCTION

The Computational Quran, developed within the discipline of Computational Religious Studies (CRS), treats revelation not as loosely associated verses but as a structured, mathematically coherent system of variables, functions, and graphs. In this framework, verses are modeled through the Five-Step Framework, mapped into the Universal Quranic Variable Ontology (UQVO), and then expressed as formal, machine-readable structures. Belief, hypocrisy, justice, charity, patience, social cohesion, and other core Quranic concepts become well-defined variables

DOI: 10.4018/979-8-3373-6940-2.ch011

with domains, relationships, and dynamics. Commands, prohibitions, promises, and warnings become functional mappings between internal states, actions, and outcomes. Narrative patterns and thematic arcs become graphs and dynamic systems. This transformation allows the Quran to be treated as an explicit epistemic architecture that can support computation, simulation, and reasoning, rather than as an opaque corpus of text.

THE RISE OF ARTIFICIAL INTELLIGENCE AND THE ROLE OF CRS

We live at a moment when intelligence is increasingly being externalized into machines. Through supervised learning, unsupervised learning, reinforcement learning, and emergent paradigms such as world models, humans are training machines to perceive, learn, generalize, and act in ways that previously required human cognition. Perception problems are being attacked with deep learning; abstraction and generalization are being strengthened through representation learning and model architectures that can transfer across tasks and domains; reasoning is being scaffolded with ontologies and knowledge graphs that encode explicit relations between entities and concepts. In short, we are not merely building tools; we are engineering artificial systems with perception, memory, inference, and decision capabilities. In such a world, the question is not whether machines will participate in human affairs, but on what terms they will do so, and according to which understanding of human nature and moral order.

If machines are increasingly embedded in decision-making, education, governance, counseling, conflict, and cooperation, it is no longer sufficient for them to understand only human work in a narrow functional sense. They must also have some structured access to what it means to be human. Human identity is not exhausted by economics, biology, or psychology. Religion remains one of the most pervasive and powerful dimensions of human life, shaping values, motivations, fears, hopes, institutions, and civilizations. To exclude religion from the computational understanding of humanity is to deliberately blind AI systems to a fundamental axis of human experience. Conversely, to include religion in a naive or purely statistical way – by merely training on religious text without structure – is to risk misrepresenting and distorting what religion actually says and does.

CRS offers a third path: it makes religious knowledge computationally accessible without flattening its complexity or discarding its theological integrity. Once the Quran is modeled through the Five-Step Framework and encoded in the UQVO, it becomes accessible not only to human scholars, who can see its internal structure more clearly, but also to artificial intelligence systems that can operate over that structure. The models produced by CRS act as an intermediate language: they translate scriptural content into variables, equations, constraints, and graphs that AI systems can ingest as world models. This does not mean teaching a machine to "believe" in a theological sense. It means teaching the machine to recognize and reason about how belief, action, intention, justice, oppression, gratitude, arrogance, and other Quranic constructs interact and evolve within human lives and societies.

In this chapter, therefore, the focus is not only on modeling the Quran for scholarly clarity, but on making CRS itself accessible to AI systems so that machines can, in a precise and disciplined way, be "trained on the Quran."

Training, in this context, means that AI systems can use Quranic models as one of their formal knowledge sources, alongside physical models, economic models, and psychological models, when reasoning about human behavior and moral choice. Mathematical formalization is what makes this possible. It allows us to couple structured scriptural models with empirical data about human behavior, social systems, and historical trends, and to study how Quranic patterns of cause-and-effect, warning and promise, rise and fall, map onto observable trajectories in the world.

Why is this important?

This integration matters because it determines whether religion will be present in the computational core of the systems that increasingly mediate human life. If revelation remains unmodeled, it remains outside the reasoning space of AI, reduced to a text that systems can quote but not structurally understand. If revelation is modeled carelessly, it becomes a caricature. But if the Quran is treated as a mathematically coherent system and encoded via CRS into AI-readable world models, then religion becomes a live source of structured insight into moral, spiritual, and behavioral dimensions of human existence. In that arrangement, AI does not stand over religion as judge, nor does religion stand outside AI as a mere critic. Instead, revelation furnishes one of the core grammars by which intelligent systems interpret and support human life.

NEW COMPUTATIONAL FOUNDATION

When sacred text is translated into structured mathematical models and then integrated with empirical data, a new computational foundation emerges for applying religious knowledge directly within human affairs. The shift is decisive: revelation is no longer confined to narrative interpretation alone but becomes part of an executable epistemic system.

Meaning moves from text to structured

Through the CRS pipeline, scriptural meaning moves from text to structured models and from structured models to AI-readable world models. In this form, religious knowledge is no longer external to computation; it becomes internal to the machinery of reasoning itself. Beliefs, actions, intentions, moral constraints, and outcomes are encoded as variables, equations, thresholds, and dynamic trajectories that can be simulated, constrained, optimized, and evaluated across time.

Improves AI systems' ability to understand religion

This transformation allows artificial intelligence systems to operate not merely on statistical correlations in religious language, but on the formal moral and spiritual logic embedded within the text. Large language models can be guided by explicit ethical structures rather than implicit training bias alone. Ethical reasoning engines can evaluate actions against revealed moral law expressed as constraints and objective functions. Spiritual advisory systems can track trajectories of belief, decay, repentance, and growth as dynamic processes rather than offering static quotations. Cognitive and societal simulations can model how faith, injustice, gratitude, corruption, patience, and power propagate through individuals and civilizations over time. In each case, mathematical structure provides the bridge that allows AI to move from superficial religious reference to deep religious reasoning.

Operationalizes religion

The deeper implication is not simply technical but civilizational. Once revelation is expressed as a coherent formal system, religion ceases to function as a black box for artificial intelligence. It becomes a structured source of moral, spiritual, and behavioral knowledge that can be reasoned over, audited, and integrated into decision systems that shape real human outcomes. In this form, religion does not compete with scientific or computational rationality; it completes it. The union of mathematical models and data does not secularize religious knowledge – it operationalizes it. It allows revealed moral architecture to participate directly in policy, governance, education, counseling, social design, and human-machine interaction at scale.

This is why the Computational Quran is not merely a digital humanities project or an exercise in symbolic abstraction. It establishes the conditions under which religious knowledge can function as an active computational force within modern technological systems. Artificial intelligence increasingly mediates human work, judgment, and social coordination. If religion remains outside that computational loop, it risks becoming ethically ornamental rather than operational. By contrast, when revelation is formally modeled and integrated into AI world models, religion reenters history not as an obstacle to technological progress, but as a governing architecture for its moral and spiritual alignment.

OPERATIONALIZING CRS IN ARTIFICIAL INTELLIGENCE

The bridge between CRS and artificial intelligence is built from three types of artifacts that together form a complete world model of Quranic logic.

Level 1: The UQVO Variables

At the most granular level are the UQVO variables. These are the canonical units of meaning – belief, hypocrisy, justice, charity, patience, gratitude, oppression, guidance, misguidance, and many others – each defined with a precise domain, structure, and set of relationships. They are not vague semantic tags. Each variable is mathematically typed, associated with tafsīr-grounded definitions, and embedded in an explicit network of causal and normative links. For an AI system, this means that "BELIEF" or "JUSTICE" is not just a word in an embedding space; it is a formally specified object with known behavior and constraints that can be tracked, updated, and reasoned over.

Level 2: Verses and Verse Clusters

On top of these variables sit the verse-level models produced by Steps 3 and 4 of the Five-Step Framework. Every modeled verse or verse-cluster becomes a small but complete system: it identifies which UQVO variables are in play, specifies how they are connected, and encodes how internal states and external actions map to outcomes. Some of these models are static functions that describe how certain conditions, when met, lead to promised rewards or warned punishments. Others take the form of dynamic systems, describing how belief, hypocrisy, gratitude, or corruption evolve over time under repeated choices. Still others are expressed as graphs that reveal the structure of conditions, exceptions, and thematic linkages. Collectively, these verse-level models form a library of micro-worlds that capture the local logic of the Quran: "if this belief and this action coexist, this trajectory or outcome is activated."

Level 3: Thematic Layer

Above this sits a global layer: large-scale thematic graphs that integrate those micro-models into a system-wide representation of the Quran's epistemic architecture. Here, nodes represent not only variables, but also verses, themes, actor-types, and composite states; edges capture causal flows, normative relations, conceptual parallels, and narrative echoes across the entire text. Belief in one context links to justice in another; patience in adversity connects to long-run spiritual elevation; oppression in a historical narrative attaches to legal prohibitions in a later passage. The result is a knowledge graph in the strong sense: a connected, navigable, and analyzable network that reveals how the Quran organizes its view of human psychology, morality, society, and destiny.

Technically, these three artifacts together give AI systems three complementary representational layers. The first is a symbolic layer: precise rules, variables, and logical constraints that define what is permissible, obligatory, forbidden, or structurally impossible within Quranic logic. The second is a graph layer: an interconnected structure of nodes and edges that captures how concepts, verses, and outcomes are interrelated, and that supports path-finding, centrality analysis, and clustering. The third is a dynamic layer: update equations, thresholds, and accumulators that describe how states change over time under different sequences of actions and conditions. When all three are available, an AI system is no longer dealing with scripture as a flat text corpus; it is interacting with a fully-fledged world model.

How AI uses these models?

From the perspective of AI design, these CRS-derived artifacts can be consumed directly as world models: structured representations of how beliefs, actions, and outcomes interact according to Quranic logic. A reasoning engine can query, "What happens to social cohesion if injustice rises and gratitude falls?" and receive an answer grounded not in mere statistics, but in explicit dynamic models linked to specific verses. A large language model can be constrained by the symbolic layer so that its ethical outputs respect the structure of obligations and prohibitions encoded in revelation. A simulation system can instantiate agents whose internal states are UQVO variables governed by Quranic dynamics, and then study how those agents behave under different conditions. This is what it means to say that CRS makes the Quran accessible to AI systems: it turns revelation into a structured world that machines can inhabit, explore, and reason within, rather than a disconnected set of quotations they merely recite.

INTEGRATION PATTERNS: HOW AI USES QURANIC STRUCTURE

Integration patterns describe the concrete ways in which CRS outputs are wired into artificial intelligence systems so that machines can actually use Quranic structure, rather than merely quote Quranic text. CRS does not stop at modeling revelation for human scholars; it produces artifacts – UQVO variables, verse-level models, and global thematic graphs – that are natively computational. These artifacts can be bound to different components of AI architectures: to language models that generate text, to reasoning engines that evaluate options, to simulators that model human behavior, and to agents that must make decisions under ethical and spiritual constraints. An integration pattern, in this sense, is an architectural template: a repeatable way of coupling CRS's formal representations to a particular class of AI system so that the system's behavior is shaped by Quranic logic, not just informed by Quranic citations.

Quran Aware AI

Once the Quran has been transformed into structured models through the Five-Step Framework and encoded in the UQVO, it becomes possible to speak meaningfully about "Quran-aware" AI. Integration patterns answer the question: what does an AI system actually do with that structure? They specify where the CRS layer sits relative to the model's generative core, how Quranic variables and rules are consulted or enforced during inference, and how the system switches between its default statistical reasoning and the alternative epistemic architecture provided by revelation. In the sections that follow, we examine several canonical patterns – beginning with symbolic constraint layers for large language models – that show, in detail, how CRS enables AI systems to treat the Quran not as a passive dataset, but as an active world model and ethical governor for their interaction with human beings.

There are several canonical integration patterns between CRS models and AI systems.

Symbolic Constraint Layer for LLMs

Large language models are powerful precisely because they are unconstrained: they generate text by extending patterns learned from vast corpora, not by obeying an explicit moral, spiritual, or epistemological calculus. Left on their own, they treat all text as statistically similar, whether it is scripture, social media, or technical manuals. The result is that when an LLM speaks about Islam or Quranic concepts, it does so in the same semantic field as ordinary discourse, even when the Quran deliberately operates with a different conceptual and epistemic architecture. The role of CRS is to insert a symbolic constraint layer between the model's raw generative capacity and its outputs on religious or other topics. In this layer, UQVO variables and Quran-derived logical rules define what is permissible, obligatory, forbidden, or inconsistent according to Quranic ethics and theology. Any answer the LLM proposes about behavior, belief, or spirituality is filtered through this structure: the system checks consistency with Quranic constraints, flags or corrects violations, and ensures that responses remain within well-defined interpretive boundaries. This turns the LLM from a free associator into a Quran-aware advisor whose creativity is disciplined by a formally modeled religious framework.

A major reason this constraint layer is necessary is that Quran-based definitions of key terms are not equivalent to how those terms appear in general language corpora. Take the concept of ʿabd. In common human languages and secular discourse, "slave," "servant," or even "worshipper" are loaded with socio-political, psychological, or metaphorical connotations that reflect human hierarchies. In the Quranic frame, ʿabd is not merely a sociological role; it is an ontological and spiritual designation describing total dependence on God, absolute ownership by the Creator, and the freedom that paradoxically comes only through servitude to the Divine rather than to creation. Similarly, taw īd is often glossed informally as "oneness" or "singularity," which invites analogies to mathematical unity or philosophical monism. But in the Quranic system, taw īd is not just numerical oneness; it is a comprehensive structure of lordship, authority, worship, and allegiance. It governs not only the count of deities but the entire arrangement of meaning, causality, and moral obligation in the universe. An unconstrained LLM, trained on mixed text, will happily conflate these Quranic notions with secular or comparative-religion notions of "monotheism," "servitude," or "unity," producing outputs that appear fluent yet are conceptually off-target.

Epistemological Change in AI

The divergence is not only semantic but epistemological. From the very beginning of the Quran's discourse – "Alif-Lām-Mīm. This is the Book about which there is no doubt, a guidance for those who are mindful of God" (2:1–2) – the epistemic frame is different from the common scientific frame that dominates much of the modern corpus used to train LLMs. Knowledge in the Quranic model is not reducible to observation and induction; it includes revelation as a primary, certain source of guidance, and it explicitly foregrounds the unseen (al-ghayb) as an object of belief and a structuring principle of reality. Human reason is affirmed but also bounded; it operates within a universe whose deepest truths are disclosed, not discovered purely by experiment. The assumptions about what counts as evidence, what counts as certainty, and how we move from premises to conclusions are therefore different from standard secular rationalism. If an LLM reasons about religious questions using only the statistical habits of secular discourse, it will unconsciously impose foreign epistemic assumptions on Quranic material, leading to subtle but profound distortion.

The constraint layer built from CRS models is designed precisely to prevent this drift. By encoding Quranic concepts as UQVO variables with tafsīr-grounded definitions, and by formalizing epistemic commitments as part of the reasoning structure rather than as mere "content," CRS forces the LLM to inhabit a different logic when it enters the Quranic world.

When the model encounters terms like ʿabd or taw īd, it is required to use the Quranic variable definitions and relations, not the looser senses circulating in general discourse. When it reasons about guidance, certainty, or the unseen, it must respect the Quran's epistemological hierarchy instead of defaulting to purely empirical criteria. In practice, this reduces hallucinations in models related to Islam and its belief systems, because the system is no longer free to invent plausible-sounding but theologically incoherent answers. Instead, it must align its generative process with an explicit, auditable structure of concepts, constraints, and permissible inferences derived from revelation itself.

In this way, the symbolic constraint layer does more than "keep the model safe." It expands the scope and depth of what the LLM can meaningfully represent. Rather than forcing Islamic concepts into the narrow channels of existing language models, CRS enlarges the model's conceptual universe by giving it access to an alternative reasoning architecture: one in which revelation, unseen realities, and divine command are not marginal anomalies but central organizing principles.

The LLM becomes capable of switching into a Quran-aware mode of reasoning that is structurally different from its default mode, treating religion not as just another topic of conversation but as a distinct, internally coherent world of meaning. That is the essence of integration: not merely constraining a general model with a few rules, but equipping it with a full, formally specified religious logic that can guide, correct, and enrich its engagement with human questions about faith, ethics, and existence.

Hybrid Neuro-Symbolic Architectures

Hybrid neuro-symbolic architectures provide a natural home for CRS because they combine the strengths of modern neural models with the precision and discipline of formal religious structure.

On the neural side, large language models, embeddings, and perception systems handle what they are best at: understanding user queries in natural language, tracking nuance, disambiguating context, and sustaining fluid interaction. They can recognize when a user is anxious, curious, hostile, or searching; they can map everyday speech into latent spaces that capture emotional tone and conversational intent. But left alone, these neural systems do not know what beliefs, patience, gratitude, or injustice mean in Quranic terms – they only know how those words behave statistically in general corpora. Quran is not meant to be a daily survey poll of behaviors.

The symbolic CRS side supplies exactly what the neural component lacks: explicit Quran-based semantics and structural logic. UQVO variables encode what concepts like BELIEF, PATIENCE, GRATITUDE, and INJUSTICE are, how they relate to each other, and how they evolve over time. Verse-level models encode structural rules such as "belief plus righteous action leads to a certain class of outcomes" or "persistent injustice leads to decay and downfall," not as slogans but as functions, thresholds, and dynamic systems.

Thematic constraints identify regions of moral impossibility – for example, that one may not justify oppression in the name of piety, or rationalize betrayal as strategic wisdom.

When the hybrid system operates, the neural component proposes interpretations, candidate answers, or plan structures, and the symbolic CRS layer evaluates, corrects, or constrains them according to Quranic logic. The result is that the AI system does not merely "know" verses as text; it operates under the structural logic those verses articulate. It can converse and adapt like a neural model while reasoning and deciding within a rigorously modeled religious world.

WORLD-MODEL INTEGRATION

In world-model integration, revelation – once expressed in mathematical form – becomes a full-blown model of human moral and spiritual dynamics that AI systems can inhabit and explore. Instead of treating Quranic content as isolated verses or abstract doctrines, the CRS framework encodes it as coupled systems of variables that evolve over time. Belief is no longer a static label but a state variable that rises or falls according to differential or difference equations driven by remembrance, sin, repentance, knowledge, and social influence. Actions are not one-off events but inputs into integrals and cumulative sums that represent the accrual of deeds, habits, and social effects over a lifetime or across generations. Societies are modeled with system dynamics: justice, gratitude, corruption, arrogance, and humility are treated as forces that shape cohesion, resilience, and decline. Thresholds in these systems mark qualitative state changes – the "sealing of hearts" after persistent rejection, the sudden moral collapse of a community after long-term corruption, the tipping points where repentance redirects a trajectory that seemed destined for ruin.

Once this Quran-based world model exists, AI systems can do much more than retrieve verses that mention a theme. They can simulate scenarios, forecast trajectories, and evaluate moral consequences within the logic of revelation itself. Given a pattern of individual behavior, an AI agent can explore how belief and spiritual state are likely to evolve under repeated actions, temptations, and interventions. Given a set of social conditions – growing injustice, concentrated wealth, erosion of gratitude – an AI simulator can examine how a community's long-run stability changes, echoing Quranic narratives of nations that were elevated or destroyed. When advising a user, a spiritually aware system can draw not only on static rules ("this is allowed, this is forbidden"), but on dynamic insights ("if you continue along this pattern, here is the kind of spiritual and social state you are moving toward").

In this mode, AI steps beyond simple "information about Islam" and begins to operate as a computational lens on the Quranic view of how human lives and societies unfold. It becomes capable of modeling moral and spiritual dynamics rather than merely describing them, which is precisely what is needed if revelation is to guide machine-mediated decision-making in a world where AI systems increasingly shape the conditions under which human beings believe, act, and live together.

APPLICATIONS

LLMs with Revelation-Aware Alignment

One of the most direct applications of CRS is a Quran-aware assistant that answers questions about life, ethics, and spirituality. From the user's perspective, this looks like a conversational system: you ask a question about a moral dilemma, a family situation, or a spiritual struggle, and the system responds in natural language with apparent fluency and empathy. Underneath, however, the interaction is governed by a tightly structured pipeline. The large language model receives the query first and does what it does best: it interprets the question linguistically and contextually, disambiguates references, and infers what the user is really asking. That raw interpretive output is then passed into the CRS layer, where it is mapped onto the relevant UQVO variables and themes – belief, patience, justice, gratitude, family roles, oppression, repentance, and so on – rather than being treated as generic "topics."

Once the query has been grounded in the ontology, the system retrieves the relevant CRS models: verse-level structures, equations, and graph segments that encode how those variables behave and interact within the Quranic framework. Reasoning now takes place not in a vacuum, but inside Quranic structural constraints. The assistant is explicitly prevented from offering advice that would violate clear prohibitions, from entertaining "ethical trade-offs" that contradict the moral architecture of revelation, or from ignoring the long-run spiritual trajectories implied by certain choices. It must consider not only whether an action is technically permissible, but also how it affects belief, character, family stability, and social justice over time according to the dynamics captured in the models.

Only after this constraint-and-reasoning phase does the LLM formulate a response in natural language, drawing on its expressive capacity but filtered, corrected, and guided by the CRS world model. The result is an assistant that is not merely "trained on Islamic text" in the superficial sense of having seen Quranic verses during pre-training; it is bound by an explicit, interpretable structure of revelation that shapes its reasoning and constrains its outputs.

ETHICAL REASONING ENGINES

Beyond conversational assistance, CRS models can power dedicated ethical reasoning engines that operate more like evaluators or judges than chat partners. These systems begin by taking a situation as input – a proposed business practice, a new regulation, a military policy, or a conflict between stakeholders. Instead of leaving that description in vague narrative form, the engine translates it into UQVO variables and relations: degrees of justice and injustice, types and levels of harm and benefit, the presence or absence of consent, the structure of intention, the use of deception, the distribution of power, and the likelihood of oppression. In other words, the scenario is recast into the formal language that the Quranic models understand. Once this representation is in place, the engine evaluates the situation against Quranic structural models: it asks, in effect, "Given how revelation describes the dynamics of oppression and fairness, harm and benefit, duty and prohibition, what does this configuration amount to?"

Concrete examples make this clearer. An engine can examine a financial or labor arrangement and determine whether it is exploitative or just by checking it against scripture-modeled variables of oppression, fairness, and consent, and against rules that forbid certain patterns of asymmetry or coercion. It can assess policies on war and peace by expressing them in terms of aggression, defense, treaty obligations, and collateral harm, and then testing those configurations against Quranic structures governing combat, protection of noncombatants, and the keeping of covenants. Technically, such an engine combines three main capabilities. First, it uses deontic logic to encode obligations, permissions, and prohibitions so that it can classify proposed actions in normative terms. Second, it uses constraint solving to check whether a proposed decision violates any relevant Quranic constraints or whether it falls within the space of permissible options. Third, it can invoke simulation to explore what a given policy is likely to do to belief, justice, social cohesion, and vulnerability to corruption over time, rather than only at a single moment. In this configuration, religion is no longer a decorative text layer that AI cites for rhetorical effect. It functions as a structural ethics engine embedded inside AI systems, providing the logic by which complex choices are evaluated and guiding decisions toward alignment with a formally modeled moral order.

SPIRITUAL ADVISORS AND COUNSEL SYSTEMS

A spiritual advisor AI built on CRS does not simply answer abstract questions about doctrine; it engages with the user's lived inner state as a modeled object. When a person describes feeling empty, overwhelmed, guilty, proud, or drifting, the system does more than match keywords. It interprets that description in terms of explicit UQVO variables: spiritual fatigue, confusion, guilt, gratitude, arrogance, despair, overconfidence, heedlessness, sincerity, and so on. These variables are not psychological labels in a generic sense; they are tightly coupled to Quranic descriptions of states of the heart and to the trajectories those states tend to follow. Once the AI has inferred a configuration of variables, it maps that configuration onto Quranic trajectories: "you are drifting into heedlessness," "you are in a phase of sincere struggle that must be protected," "your gratitude is growing but pride is beginning to appear," "your despair is obscuring the doors of repentance." The system is not diagnosing pathology; it is locating the user within a Quranic map of spiritual motion.

From that mapped position, the advisor can recommend interventions that are consistent with Quranic structure and modeled dynamics. For someone trending toward heedlessness and moral numbness, it may highlight remembrance, gratitude, and small consistent acts of charity. For someone struggling sincerely but discouraged, it may emphasize patience, reflection on specific narratives of prophetic endurance, and constructive consultation with trusted human scholars and community members. Because the CRS model includes dynamic equations for how belief changes, accumulation structures for how deeds and habits build up over time, and thresholds where qualitative shifts occur (such as the hardening or softening of the heart), the advisor can talk not only about what is right in principle but also about how the person is moving through spiritual space. Its guidance can therefore be consistent across time, traceable back to specific Quranic models and variables, and theologically grounded rather than ad hoc or purely therapeutic.

Crucially, this kind of system is not an "AI muftī" replacing scholars or issuing binding legal rulings. It is better understood as a structured reflector of Quranic spiritual dynamics: a tool that helps individuals see themselves through the lens of revelation's modeled patterns and that continually points back to human scholarship, community, and worship as essential contexts. Its authority is not independent; it is derivative and subordinate, confined to what can be formally justified within the

CRS framework. In that limited but important role, a CRS-based spiritual advisor can scale compassionate, theologically coherent reminders and reflections to many people, while preserving the centrality of human scholars and the lived tradition in guiding the deeper and more complex questions of religious life.

COGNITIVE AND SOCIETAL SIMULATION MODELS

CRS models also open the door to a class of applications that are neither purely conversational nor purely evaluative, but explicitly exploratory: simulations of individual and collective life as seen through the Quranic lens. At the individual level, CRS enables micro-simulations of cognitive and spiritual journeys. Each simulated person can be represented as an agent with UQVO-based internal states – belief, sincerity, hypocrisy, patience, gratitude, greed, fear, hope – and behavior variables such as charity, injustice, remembrance, and transgression. Dynamic equations govern how those states change in response to habits, trials, social environments, and interventions. One can then ask: how does a person's belief and moral character evolve over time if they persist in a certain pattern of action, are exposed to particular influences, or respond to hardship in different ways? The simulation does not claim to capture the full mystery of human souls, but it formalizes the structural patterns that the Quran repeatedly describes: that certain behaviors harden or soften the heart, that repeated injustice corrodes spiritual integrity, that gratitude draws further blessing, and that sincere repentance redirects trajectories that appeared lost.

At the societal level, CRS supports macro-simulations in which communities are modeled not as single homogeneous actors but as dynamic systems with aggregate variables: social cohesion, corruption, justice distribution, gratitude, arrogance, vulnerability to internal decay, and resilience against external shock. Here, differential or difference equations encode how increases in injustice, concentration of wealth, erosion of trust, and abandonment of shared moral norms affect the long-run stability of a society.

The Quranic narratives of nations that rose and fell – because of arrogance, denial, oppression, or neglect of the vulnerable – are translated into system dynamics that can be explored computationally. One can test scenarios such as: what happens to a community's cohesion if its economic indicators improve but its moral norms decay? How does the long-term trajectory differ if justice is prioritized early versus deferred? These simulations do not replace historical analysis or policy judgment; they provide a rigorous way to explore the Quran's own causal motifs in a form that can be probed, varied, and visualized.

The outcome of such work is twofold. First, it allows Quranic patterns of the rise and fall of individuals and nations to be expressed as formal dynamic systems that can be studied in the same analytical space as economic or ecological models. Second, it creates the basis for policy-support tools that incorporate moral and spiritual trajectories as first-class variables alongside economics, security, and technology. A government decision-support system informed by CRS would not only estimate GDP, employment, or defense readiness; it would also estimate what a given policy is likely to do to justice, corruption, social trust, and the moral direction of the populace. In this way, the Quran's view of human and societal dynamics can meaningfully inform deliberation about the future, not as a rhetorical afterthought but as a computationally integrated dimension of strategic planning.

TECHNICAL PLACEMENT: WHERE CRS SITS IN THE AI STACK

In architectural terms, CRS functions as a structured intermediary layer between the raw world of data and text on one side, and the visible behavior and decisions of AI systems on the other. At the base of the stack are the sources themselves: Quranic text, tafsīr, historical context, and ongoing user interactions. Taken alone, this material is just that – data – rich, sacred, and complex, but not yet in a form that machines can reason with in a disciplined way. The CRS layer sits directly above this foundation. It applies the Five-Step Modeling Framework, constructs the Universal Quranic Variable Ontology (UQVO), and builds verse-level models, dynamic systems, and thematic graphs. In doing so, it transforms raw text and commentary into formal objects: variables with defined domains, equations describing trajectories, graphs encoding conceptual relationships, and logical rules specifying obligations, prohibitions, and constraints.

Above the CRS layer sits what we can call the world model layer. Here, the models produced by CRS are assembled into an AI-readable representation of the Quranic universe: a belief space, an action space, and an outcome space, all linked by ethical constraints, spiritual trajectories, and societal dynamics. This layer is where the Quran's view of how beliefs, actions, and consequences interrelate is encoded as a coherent system that can be queried, simulated, and updated. It is the living map of Quranic logic that AI agents consult. Those agents – large language models, reasoning engines, advisory systems, and simulation frameworks – occupy the next layer. They interact with users, generate text, propose decisions, and run simulations, but they do so by querying the CRS-informed world model, being constrained by its logic, and using it as a reference point for evaluation and decision-making. In this design, AI behavior is not free-floating; it is anchored in an explicit scriptural structure that can be inspected and improved.

The highest layer is the human interaction layer, where scholars, users, policy-makers, and communities encounter the outputs of these systems. They read, critique, validate, and govern how CRS-informed AI is used and how its models are updated. The presence of this layer is essential: it ensures theological accountability, contextual judgment, and communal ownership of how revelation is operationalized. Together, these layers define CRS as a form of middleware that translates divine guidance into a formal language AI can operate with, without severing that guidance from human oversight. It mediates between Revelation and machine, preserving the integrity of both: the Quran retains its status as the ultimate source of meaning, while AI gains a principled, auditable way to engage with that meaning in the service of human life.

GOVERNANCE, TRANSPARENCY, AND THEOLOGICAL ACCOUNTABILITY

Because CRS is formal and explicit, it is intrinsically auditable in a way that conventional "AI trained on religious text" is not. Every AI decision or recommendation grounded in CRS can, in principle, be traced back through a chain of artifacts: to specific UQVO variables that define the concepts in play, to specific verse-level models that encode the relationships between beliefs, actions, and outcomes, and to specific tafsīr-based definitions that anchor those models in the interpretive tradition. Instead of a system that says, "the model thinks this is Islamic," you have a system that can say, "this recommendation arises because these variables crossed these thresholds according to this model, which is derived from these verses interpreted in this way." The logic is no longer buried in opaque weights; it is exposed in structured revelation.

This traceability is what makes scholarly oversight and real governance possible. Scholars can examine the ontology entries and verse models, review whether the tafsīr grounding is appropriate, and refine structures or parameters when they judge that the model has oversimplified or misrepresented the tradition. Changes become versioned events rather than invisible adjustments: a variable definition is revised, a threshold is moved, a dynamic equation is updated, and each such change is documented, reasoned, and attributable. Over time, the CRS framework accrues a history of interpretive decisions that can itself be studied and debated, instead of silently drifting as the model retrains on new data.

For end-users and communities, this yields a kind of transparency that is largely absent from current religiously themed AI systems. A generic model "trained on Islamic content" cannot explain why it answered a legal or ethical question a certain way; its justification is, at best, a post hoc rationalization over a hidden statistical process. A CRS-driven system, by contrast, can show why a recommendation was made in terms of structured revelation: which Quranic models were invoked, which constraints were binding, and how the user's described situation mapped to the relevant variables. This does not make the system infallible – its models can still be incomplete or contestable – but it makes its operation intelligible and open to correction. Theological accountability is thus built into the architecture: the community can see, understand, and, when necessary, challenge the way revelation has been formalized and applied.

REFRAMING RELIGION AND AI

Conventional narratives often treat religion as the opposite of what AI needs: irrational, unstructured, incompatible with formal methods, and ultimately a barrier to technological rationality. In that story, science and computation belong to one universe, while revelation belongs to another; at best, religion is tolerated as private meaning, at worst it is seen as noise to be excluded from serious systems. The Computational Quran and CRS overturn this assumption by changing the *form* in which religion presents itself to machines. When Quranic content is expressed as variables with defined domains, equations that describe how states and behaviors evolve, graphs that encode conceptual and thematic structure, and logical constraints that define obligations, prohibitions, and impossibilities, religion no longer appears as a loose collection of narratives and injunctions. It reveals itself as an organized epistemic system, a structured moral calculus, and a dynamic model of human and societal behavior operating under its own coherent logic.

Seen in this light, religion is not in conflict with AI; it is indispensable for any AI system that claims to understand or support human beings in their full moral, spiritual, and behavioral depth. A purely secular world model can track incentives and preferences, but it cannot natively represent repentance, divine accountability, unseen realities, or the way revealed guidance reshapes identity, motivation, and obligation. CRS closes that gap by repositioning revelation from "data to train on" to *logic to build with*. The Quran is no longer just another text in the training corpus; it becomes a structured, formal, living framework that can govern how an AI system reasons about good and evil, harm and benefit, justice and oppression, hope and despair. Not a relic, not a dataset, but a foundational architecture that should play a central role in the design of ethically grounded, spiritually aware artificial intelligence.

HOW THE SYMBOLIC CONSTRAINT LAYER INTERFACES WITH THE NEURAL GENERATIVE CORE

Although the long-term objective of this work is to support the next generation of artificial cognition, especially world models, the immediate technical challenge is more practical: how can structured symbolic meaning be connected to today's neural systems, including large language models? This section therefore treats RAG, knowledge graphs, and symbolic constraints not as the final architecture, but as transitional mechanisms that allow Quranic structure to be grounded, represented, and interfaced with current AI systems on the path toward world-model-based cognition.

The simplest way to explain the architecture is this: the neural core generates possibilities, while the symbolic layer governs admissibility, structure, and interpretive discipline. The neural model is good at producing fluent language, detecting patterns, retrieving associations, and generalizing across examples. The symbolic layer is responsible for representing explicit Quranic concepts, relationships, rules, constraints, interpretive assumptions, and moral boundaries. In a Computational Quran system, the neural model should not be allowed to generate religious interpretations entirely from statistical association. Its outputs should be shaped, checked, and, when necessary, blocked or revised by the symbolic layer.

Technically, this interface can operate at four levels.

First, the symbolic layer can operate before generation through structured retrieval. When a user asks a question or when the system analyzes a verse, the query is not sent directly to the neural model alone. It is first mapped to a structured representation: relevant verses, concepts, agents, actions, moral states, legal categories, theological principles, and prior model variables. These may be stored in a knowledge graph or ontology. The system then retrieves the relevant structured context and passes it to the neural model as grounded input. This is similar to retrieval-augmented generation, where a model combines parametric knowledge with retrieved non-parametric memory to improve factuality and specificity. Lewis et al. showed that retrieval-augmented generation improves knowledge-intensive NLP tasks by giving language models access to retrieved evidence rather than relying only on internal parameters (Lewis et al., 2020).

Second, the symbolic layer can operate during generation through constrained decoding. Normally, an LLM predicts the next token from a probability distribution. But a constraint layer can modify this distribution before the next token is chosen. Tokens, phrases, or structures that violate the symbolic rules can be down-weighted, blocked, or forced into a required format. For example, if the system is generating a model of repentance, it may be required to include variables such as recognition, regret, timing, sincerity, and behavioral correction. If it is producing an interpretive claim, it may be required to state the relevant verse cluster, governing principle, scope condition, and uncertainty level. Research on grammar-constrained decoding shows that formal constraints can improve syntactic correctness and semantic accuracy in logical parsing tasks, especially when LLMs are used to translate natural language into symbolic forms. (Raspanti et al., 2025).

Third, the symbolic layer can operate after generation through verification and critique. In this mode, the neural model first proposes an interpretation, equation, variable map, or moral-state transition. The symbolic layer then checks the output against the Quranic knowledge graph, known constraints, tafsir references, variable dictionary, and formal consistency rules. If the output violates a rule, omits a required variable, contradicts a verse cluster, or makes an unsupported claim, the verifier returns an error signal. The system then revises the answer. This creates a generate-check-revise loop. Neuro-symbolic AI research broadly supports this architecture because it combines the pattern-recognition strength of neural networks with the explicit reasoning and knowledge-representation strength of symbolic systems. DeLong et al. describe neuro-symbolic AI as a field that combines symbolic reasoning with deep learning to leverage their complementary benefits, especially for reasoning over knowledge graphs (DeLong et al. 2025).

Fourth, the symbolic layer can operate during learning or fine-tuning by converting symbolic constraints into training signals. In this version, the model is not merely checked after output; it is trained to internalize the symbolic structure. The system can be given examples of valid and invalid interpretations, correct and incorrect variable mappings, acceptable and unacceptable state-transition models, and revised models after failed simulations. These examples become supervised, contrastive, or reinforcement-style feedback. Over time, the neural core learns to generate outputs that are more consistent with the symbolic layer. Recent neuro-symbolic surveys identify representation, learning, reasoning, and decision-making as central features of neuro-symbolic systems, which is precisely the combination needed for a Computational Quran architecture (Bhuyan et al., 2024; Besold et al., 2017).

A useful way to visualize the interface is as a pipeline:

[Text→ Verse/ConceptParser→ QuranicKnowledgedGraph→ SymbolicConstraintLayer→ NeuralGenerator→ Verifier→ RevisedOutput]

The neural generator is not removed. It remains essential because religious language, historical context, analogy, and explanation require flexible language generation. But the generator is no longer sovereign. It operates inside a structured interpretive environment.

For example, suppose the system is asked to model repentance. The neural core may propose:

[Repentance = Recognition + Regret + BehavioralCorrection]

The symbolic layer then tests this model against relevant Quranic cases, including Pharaoh's late declaration of belief at the moment of drowning. The verifier detects that the model misclassifies compelled last-moment recognition as valid repentance. It therefore identifies missing variables: timing, sincerity, and opportunity for moral transformation. The revised model becomes:

[RepentanceValidity = f(Recognition, Regret, BehavioralCorrection, Timing, Sincerity, PreDeathOpportunity)]

This example shows the interface in action. The neural core proposes a plausible model. The symbolic layer tests it against structured scriptural knowledge. The verifier finds a failure. The system revises the model. The result is not merely a better answer; it is a more disciplined interpretive process.

In the Computational Quran, the symbolic constraint layer therefore serves three purposes. It grounds generation in the text. It prevents unsupported or structurally invalid interpretations. And it makes reasoning reproducible by forcing the system to expose which variables, rules, assumptions, and verse clusters shaped the output. This is the difference between an AI that merely talks about the Quran and an AI system that reasons within a Quranic model of meaning.

REFERENCES

Besold, T. R., d'Avila Garcez, A., Bader, S., Bowman, H., Domingos, P., Hitzler, P., Kühnberger, K.-U., Lamb, L. C., Lowd, D., Lima, P. M. V., de Penning, L., Pinkas, G., Poon, H., & Zaverucha, G. (2017). *Neural-symbolic learning and reasoning: A survey and interpretation*. arXiv. /arXiv.1711.03902DOI: 10.48550

Bhuyan, B. P., Ramdane-Cherif, A., Tomar, R., & Singh, T. P. (2024). Neuro-symbolic artificial intelligence: A survey. *Neural Computing & Applications*, *36*(21), 12809–12844. DOI: 10.1007/s00521-024-09960-z

DeLong, L. N., Fernández Mir, R., & Fleuriot, J. D. (2025). Neurosymbolic AI for reasoning over knowledge graphs: A survey. *IEEE Transactions on Pattern Analysis and Machine Intelligence*. PMID: 39024082

Lewis, P., Perez, E., Piktus, A., Petroni, F., Karpukhin, V., Goyal, N., & Kiela, D. (2020). Retrieval-augmented generation for knowledge-intensive nlp tasks. *Advances in Neural Information Processing Systems*, *33*, 9459–9474.

Raspanti, F., Beurer-Kellner, L., Fischer, M., & Vechev, M. (2025). Grammar-constrained decoding makes large language models better logical parsers. In *Proceedings of the 63rd Annual Meeting of the Association for Computational Linguistics: Industry Track*. Association for Computational Linguistics. DOI: 10.18653/v1/2025.acl-industry.34

Chapter 12
CRS and Interpreting Scriptures

ABSTRACT

Chapter 12 argues that the central problem of scripture is not revelation itself but interpretation – often subjective, inconsistent, and shaped by unspoken assumptions. Computational Religious Studies (CRS) introduces a new epistemic layer that formalizes interpretive claims as explicit models, making variables, relationships, and dynamics transparent, reproducible, and testable. By converting readings into structured hypotheses grounded in the Universal Quranic Variable Ontology, CRS exposes hidden premises, enables comparison across schools of thought, and supports simulations of social and moral outcomes. It preserves legitimate pluralism while reducing ambiguity, equipping scholars with rigorous tools and ensuring theological accountability. In doing so, CRS reframes hermeneutics as disciplined model-building that aligns sacred meaning with a computational age.

INTRODUCTION

Interpretation has always been the central problem of scripture. From the earliest generations to the present, sincere readers have disagreed about what verses mean, how they apply, and which principles should govern their use in law, politics, and personal life. These disagreements are not merely theoretical. They have fractured communities, hardened sectarian identities, and, in some cases, given religious cover to oppression, patriarchy, authoritarianism, and violent extremism. The same text has been invoked to justify radically different projects: liberation and empire, justice and domination, mercy and cruelty. Behind this lies a simple reality: interpretation is often subjective and contested, even when interpreters share the same sacred words. They lack a shared, explicit methodological ground on which interpretive

DOI: 10.4018/979-8-3373-6940-2.ch012

claims can be compared, examined, and, when necessary, challenged at the level of structure rather than rhetoric.

ENVISIONING APPLIED CRS

The classical disciplines of tafsīr, u ūl al-fiqh, and kalām represent some of the most sophisticated intellectual achievements of the Islamic tradition. They provide genealogies of interpretation, tools for weighing evidence, principles for legal derivation, and theological frameworks for understanding attributes, causality, and moral responsibility. Yet for all their depth, they remain largely textual and discursive. Interpretations are presented as arguments, not as formal models. A scholar may reason in a highly disciplined way, but the model in his or her mind – how variables like intention, harm, benefit, justice, and public interest interact – is rarely made explicit in a way that others can systematically test or reproduce. As a result, interpretive outcomes are often non-reproducible, depending heavily on the authority, training, and intuition of individual scholars. They are non-falsifiable, because there is no shared structural representation that can be confronted with data or with alternative models. And they are non-transparent, in the sense that assumptions about human nature, social order, causality, and epistemology operate in the background rather than being formally articulated and open to scrutiny.

This is where Computational Religious Studies (CRS) proposes an epistemic shift. CRS does not claim to replace tafsīr or the classical sciences; rather, it introduces a new layer that formalizes their structural outputs. When a scholar interprets a verse about justice, family, governance, or war, CRS asks: what variables are being assumed, what relationships between them are being claimed, what dynamics over time are being implied? It then encodes those assumptions and claims as explicit models: variable definitions, functions, thresholds, logical constraints, and dynamic systems. An interpretive position becomes a model hypothesis. It can be reconstructed step by step, compared with alternative models derived from different readings, and examined for internal coherence and compatibility with the broader Quranic structure.

The central thesis is that CRS converts interpretive claims into reproducible, transparent, and analyzable models, thereby providing a more stable epistemology for sacred meaning. Reproducibility arises because, once the formal recipe for a given interpretation is written down, any qualified reader can rebuild the model and see whether it behaves as claimed. Transparency follows from the fact that all the hidden background assumptions – about what counts as justice, what constitutes harm, how intention modifies moral status – must be spelled out in the model rather than smuggled in rhetorically. Falsifiability becomes possible, not in the crude sense of "disproving" revelation, but in the more precise sense of testing whether a particular interpretive model fits the structural and empirical realities it claims to describe. In this way, CRS does not close the door on hermeneutics; it frames interpretive work within a discipline that makes it more honest, comparable, and accountable to both the text and the world it addresses.

WHAT "INTERPRETATION" HAS HISTORICALLY MEANT

When we speak about "interpretation" in the history of Islam, we are referring to a landscape that is intrinsically plural. From early on, the ummah organized itself into schools of law, theological currents, and regional intellectual traditions – each with its own sensibilities and methods. The madhāhib developed distinct approaches to analogy, public interest, and textual evidence. Theologians differed on divine attributes, human freedom, and causality. Later movements debated the nature of political authority, the conditions for legitimate jihad, the status of women, and what it means for a state to be "Islamic." These disagreements are not marginal footnotes; they shape law, institutions, and everyday life. Interpretive pluralism is, in one sense, a sign of vitality and depth. Yet it also demonstrates that "interpretation" has meant radically different things to different communities, even when they stand before the same verses.

Part of this pluralism comes from structural sources of subjectivity built into the interpretive process. The Arabic of the Qur'an is dense with meaning: words are polysemous, metaphors are layered, and ellipsis invites completion from context. Historical distance adds another layer of uncertainty; we possess only partial records of circumstances of revelation, early practice, and social conditions. Beyond language and history, interpreters are human. They bring cognitive biases, political interests, and socio-economic assumptions to their reading, often unconsciously. A jurist working under a caliphate, a scholar in a colonial context, and a modern intellectual in a nation-state will not approach issues of authority, resistance, or public order in the same way. Beneath all of this lies something more subtle: every serious interpreter operates with an "unspoken model" in mind – a set of assumptions about how individuals, families, markets, and states work, how harm and benefit flow, what motivates people, and what a healthy society looks like. This model silently drives exegesis but is rarely spelled out, let alone examined as an explicit object of analysis.

The result is a persistent hermeneutic gap between text, interpretation, and application. On paper, the movement looks linear: verses are read, interpreted, and then translated into systems of law, governance, ethics, and policy. In practice, there is no standard way to move from one layer to the next, and no shared formal language in which different interpretive paths can be compared. One scholar's reasoning about political authority or gender roles may be compelling to those who share his assumptions and training, but opaque or unconvincing to those outside his school. When interpretations clash, the debate often devolves into rhetorical contest, appeal to traditional loyalties, or raw institutional power. There is no agreed-upon structural representation that would allow us to say: "This reading produces these social dynamics and rests on these assumptions; that reading produces different dynamics and rests on others." Without such a representation, interpretive disputes cannot be tested against their own claimed logic, let alone against data about how real societies behave when those interpretations are applied. This is the hermeneutic gap that CRS seeks to address, not by eliminating disagreement, but by giving it a formal arena in which its stakes, assumptions, and consequences can be made explicit.

CRS proposes an epistemic shift in how we understand interpretation itself. In the classical mode, exegesis is largely an exercise in careful reading, linguistic analysis, cross-referencing verses and hadith, and reasoning within the principles of u ūl al-fiqh and theology. The outcome is a reasoned opinion, often supported by chains of argument and citation. In CRS, that process is not rejected, but extended. The Five-Step Framework and the Universal Quranic Variable Ontology (UQVO) act as interpretive formalizers: once a verse has been read and explained, the interpreter asks what variables, relations, and dynamics are being implicitly assumed. Those elements are then made explicit and encoded as equations, graphs, and logical constraints. Interpretation, in other words, becomes model construction. The question is no longer only "What does this verse mean?" but also "What are the formal structures that this meaning presupposes?"

Seen this way, every serious interpretive claim corresponds to a model hypothesis. When someone says "Justice in the Quran means X," that is not just a semantic statement; it implies a particular structure of variables and thresholds: who counts as a rights-holder, what counts as harm or redress, how resources and responsibilities should be distributed, and what outcomes are considered acceptable or oppressive over time. When a scholar or movement claims "An Islamic state must do Y," that too implies measurable parameters for governance, economic policy, social welfare, and accountability. Under CRS, such claims are not left at the level of rhetoric. They are translated into explicit models: networks of UQVO variables, functional relationships that map choices to consequences, and dynamic systems that describe how societies evolve under the proposed arrangement.

Once interpretive claims are cast in this formal light, they become available for computation. Instead of remaining as commentaries that can be read, admired, or contested only through more commentary, these models can be simulated. One can ask: if we adopt this definition of justice and this structure of an Islamic state, what happens to different segments of the population over time under plausible conditions? Do the patterns that emerge resemble the Quranic descriptions of flourishing, decline, or divine censure? Models can be stress-tested with data: corruption indices, inequality measures, family stability metrics, and conflict statistics can be examined inside the Quranic structural frame to see whether an interpretation claiming to embody justice actually produces trajectories aligned with Quranic promises and warnings.

Most importantly, interpretive outputs can now be compared side by side as alternative formal models rather than as competing opinion clusters. Two schools may differ on how a particular verse about governance should be applied. Under CRS, each school's position yields a distinct model: different variables highlighted, different thresholds set, different feedback loops assumed. Those models can be analyzed for internal coherence, cross-checked against the rest of the Quranic structure, and evaluated against historical or contemporary data. Disagreement does not disappear, but it moves from an opaque clash of authorities to a structured comparison of explicit hypotheses. In this shift from opinion to formal models lies the epistemic promise of CRS: it does not claim to deliver final meaning, but it does insist that interpretive claims be made precise enough to be examined, tested, and, if necessary, revised in the light of both revelation and reality.

Mathematical formalization in CRS is often misunderstood as an attempt to "freeze" or exhaust the meaning of revelation. It is essential to state clearly what it does not do. It does not seal meaning or pretend that a set of equations can capture the full depth of a divine text that has guided hearts and civilizations for centuries. It does not render tafsīr, historical inquiry, or the classical sciences obsolete; on the contrary, it depends on them for its inputs. Nor does it reduce revelation to mere equations, as if the Quran were just a technical manual waiting to be translated into symbols. The purpose of formalization is not to replace the rich, multi-layered discourse of the tradition, but to give that discourse a structural expression that can be analyzed, compared, and put into disciplined conversation with data and systems thinking.

The role of mathematics in CRS is better understood as scaffolding than as closure. It provides a framework within which meaning can develop and be evaluated, without dictating the meaning itself. When we formalize an interpretive position, we are forced to define variables clearly: what, exactly, do we mean by justice, oppression, guidance, hypocrisy, gratitude, or social cohesion? We are compelled to make relationships explicit: if we say that injustice leads to social decay, in what way, through which channels, and under what conditions? We must specify dynamics: how do belief and behavior evolve over time under repeated choices, and what thresholds mark qualitative changes such as moral collapse or renewal? We also have to articulate assumptions and boundary conditions: what is being held constant, what is allowed to change, and what counts as evidence of alignment or misalignment with Quranic patterns. Semantics, history, and contextual analysis continue to supply the content – the texture of meaning, the narrative, the situated understanding of verses – while mathematics supplies the form in which that content is structurally organized.

Rigor matters because, once interpretive content is encoded in this formal structure, it becomes something that others can reproduce, scrutinize, and test. Reproducibility means that another scholar can take the same verses, the same tafsīr, and the same modeling rules and arrive at the same structural representation, or else identify precisely where and why they diverge. Scrutinizability means that assumptions are no longer buried in prose; they are visible in the choice of variables, the shape of functions, the placement of thresholds, and the topology of graphs. One can ask: Why was this variable omitted? Why was this effect treated as linear rather than nonlinear? Why was this feedback loop ignored? Testability means that models can be confronted both with the broader Quranic corpus and with empirical data. If an interpretation claims that a certain political or economic arrangement is "Islamic" and conducive to flourishing, the corresponding model can be run against real-world data about corruption, inequality, family breakdown, violence, and trust to see whether the trajectories it predicts match the trajectories we actually observe. In this way, mathematics does not end hermeneutics; it disciplines it. It creates a space where sacred meaning is pursued with the same seriousness of structure and evidence that we expect in other domains of knowledge, while still acknowledging that revelation transcends any single model we may build of it.

One of the most powerful interventions CRS makes is at the level of semantics itself. In ordinary AI practice, meanings are inferred from how words are used across vast textual corpora; "meaning" is essentially the statistical behavior of tokens in context. The Universal Quranic Variable Ontology (UQVO) reverses that logic for scriptural concepts. It does not ask, "How does the word ʿabd behave in modern Arabic or English corpora?" It asks, "What does ʿabd mean within the Quranic world, as established by the text and its tafsīr tradition?" In a general corpus, ʿabd may be glossed as "slave" or "servant," carrying historical baggage of human ownership, exploitation, and social subordination. In the Quranic frame, however, ʿabd is an ontological category: the human being as utterly dependent on God, owned by Him alone, and liberated from servitude to anything else precisely through exclusive servitude to Him. UQVO encodes this as a structured variable linked to concepts of lordship, accountability, and freedom from shirk, not as a generic role in a social hierarchy.

The same holds for taw īd. In ordinary discourse, it is often reduced to the idea of "oneness" or "monotheism," inviting analogies to numerical singularity or philosophical notions of unity. Within CRS, taw īd is modeled not as a bare count but as a complex structure: the unification of worship, allegiance, love, fear, hope, and obedience in relation to one Lord, and the ordering of causal explanation such that ultimate causality is attributed to God even as proximate causes operate in the created world. Likewise, ʿadl (justice) cannot be left as a vague moral ideal. In UQVO it becomes a precise construct involving the balance of rights and duties, the fair distribution of burdens and benefits, transparent accountability, and alignment with divine measurement (mīzān). This structured semantics forces any interpretive or computational system that uses these variables to respect Quranic meanings, rather than unconsciously importing secular, colonial, or purely modern political connotations under the same labels.

CRS goes further by encoding not only semantic content but also the Quran's epistemological frame. The opening of Sūrat al-Baqarah – "This is the Book about which there is no doubt, a guidance for the God-conscious" – marks a starting point that is fundamentally different from that of secular science. Revelation is presented as a primary, certain source of knowledge, not as one data point among many. Belief in the unseen (al-ghayb) is not a residual category of ignorance; it is a central epistemic commitment that shapes how evidence is weighed and how reality is conceived. In the CRS framework, this is not treated as a pious preamble and then ignored. It is baked into the models: variables representing revelation, unseen realities, divine promise and threat, and eschatological outcomes are treated as legitimate components of the world model. This means that when CRS-based systems reason about human life, they do so in a space where guidance, divine oversight, and the afterlife are structurally present, not as rhetorical embellishments but as operational variables.

Closely related is the way CRS turns reasoning assumptions themselves into modeled objects. In conventional discourse – and in LLM-style "reasoning" in particular – assumptions about what counts as evidence, how certainty is graded, and which causal explanations are acceptable usually go unexamined. The model simply mirrors the epistemic habits of the text it has seen most, which in practice means a largely secular, empiricist, and presentist frame. CRS makes these assumptions explicit. It demands that we specify what sources are treated as axiomatic (revelation, sound reports), what is treated as probabilistic but respectable (empirical data, historical patterns), and what is considered conjectural. It requires us to categorize degrees of certainty (yaqīn, ann, wahm) and to formalize which causal claims are allowed or disallowed within the Quranic world model – for example, denying that events are purely random or that material causes exhaust explanation.

Because these epistemic commitments are coded into the structure of models, a CRS-informed system does not slip, by default, into the secular assumptions of its training corpus. It cannot casually treat revelation as "just another opinion," or bracket the unseen as irrelevant to serious reasoning. It must reason within the Quranic architecture of knowledge, causes, and purposes that has been formally specified. In this way, CRS does not only correct the meanings of individual terms; it relocates hermeneutics inside a different universe of discourse, one in which what counts as real, knowable, and important is defined by revelation itself and made operable for analysis and computation.

EXTRACTING IDEAL STRUCTURES: CIVILIZATION, INSTITUTIONS, AND POLICY

CRS does not stop at the level of individual verses or personal ethics; it opens the door to asking a much larger and more demanding question: what would an ideal civilization look like in Quranic terms? Once revelation has been translated into variables, relationships, and dynamic systems, it becomes possible to treat "ideal civilization" not as a slogan, but as a modeled target. High-level models can be constructed in which justice, knowledge, worship, economy, family, governance, and environment are not separate silos but interacting subsystems. Each subsystem is linked to specific Quranic principles – fair distribution of wealth, protection of the vulnerable, honoring family ties, truthfulness in contracts, stewardship of the earth, consultation in governance – and each has its own internal dynamics and thresholds. The question "What is a Quranically sound civilization?" can then be formalized as "What configuration and balance of these subsystems yields stability, justice, and spiritual flourishing according to the structural patterns of the Quran?"

From there, CRS enables a disciplined movement from verses to institutional blueprints. Verse-level models that describe, for example, how justice must be upheld in testimony, how wealth should circulate, how disputes should be resolved, or how power should be constrained can be mapped onto institutional variables: courts and their independence, market rules and transparency, welfare systems and safety nets, educational curricula and access, mechanisms of conflict resolution, and orientations in foreign policy. Rather than reading these verses as purely moral exhortations, CRS expresses them as requirements on system design. System-dynamics tools can then be used to explore which combinations of institutional arrangements produce stable, Quran-consistent trajectories and which lead to patterns of corruption, breakdown, or oppression that the text warns against. This moves the conversation about "Islamic institutions" from abstract ideals and historical nostalgia into a space where configurations can be explicitly proposed, modeled, and assessed.

The same logic applies to policy design. Much of what passes as "Islamic" policy today is ad hoc: a modern policy is drafted and then justified post hoc with selected verses or hadith, or a vaguely defined notion of "Islamic values" is invoked without a clear structural link to the Quran's own models of social and moral dynamics. CRS offers a different route. It provides normative-structural blueprints derived from the Quran's modeled logic and uses them as a constraint space within which policies can be crafted. A candidate policy on taxation, urban development, media regulation, or international alliances can be encoded into the model: how does it affect variables such as justice, concentration of wealth, protection of dignity, family cohesion, or vulnerability to oppression over time? Does it move the system toward or away from the stable regions that Quranic dynamics identify as desirable? If it claims to be "Islamic," can that claim be defended within a CRS model, or does the simulation reveal hidden contradictions with the Quran's structural vision?

In this way, CRS does not pretend to automate governance or replace human judgment. Rather, it offers a disciplined environment in which visions of Islamic civilization, institutional design, and policy can be articulated, tested, and refined. It makes it harder to hide behind pious labels while enacting structures that reproduce injustice or moral decay. And it allows scholars, policymakers, and communities to think together about the long-term shape of a Quranically grounded civilization with a level of structural clarity that was previously unavailable – linking text, interpretation, institutions, and outcomes into one continuous, analyzable system.

REDUCING AMBIGUITY, RESPECTING DIVERSITY

CRS must be understood first as a clarifier, not a dictator of meaning. Its primary function is to reduce unnecessary ambiguity by forcing assumptions into the open. When a scholar or school proposes an interpretation, CRS asks: what variables are you assuming, what causal relationships do you believe the Quran is asserting, what thresholds or exceptions do you think apply? By encoding those elements formally, CRS makes explicit what is usually implicit. This alone resolves a great deal of confusion, because many disagreements turn out not to be about the verses themselves, but about hidden premises regarding human nature, social order, or epistemic priorities. At the same time, CRS is not designed to impose a single "final" interpretation where the tradition itself recognizes valid multiplicity. Its rigor targets vagueness and incoherence, not legitimate pluralism. Where the Quran or the classical sciences have left room for more than one structured reading, CRS allows that multiplicity to stand – only now each reading appears as a clear model rather than a vague label.

A crucial advantage of formalization is that multiple interpretive traditions can be encoded side by side without being flattened into a single hybrid. If Hanafi jurists and Shafi'i jurists weight certain evidences differently or reason differently about public interest, this can be captured as Model A and Model B, each with its own UQVO variable emphases, relational structures, and decision rules. Likewise, differing theological positions on issues like predestination, causality, or divine attributes can be represented as alternative model families with distinct assumptions about what counts as an acceptable causal explanation or how human choice interacts with divine decree. CRS thus preserves diversity by showing differences structurally rather than collapsing them into a generic "Islamic position." Instead of glossing over divergence in the name of unity, it gives each tradition a precise computational profile, making it possible to understand both where they genuinely converge and where they part ways.

Once interpretations are represented as formal models, they can be evaluated comparatively in a way that is impossible when they remain only as prose. Competing models can be examined for internal coherence: do their assumptions contradict each other, do their dynamics lead to outcomes that the model itself claims are unacceptable, do they silently rely on principles they officially deny? They can be checked for consistency with the broader Quranic structure: does a particular reading of a verse on governance or gender harmonize with the Quran's wider treatment of justice, mercy, and human dignity, or does it create tensions elsewhere in the corpus? Models can also be tested for empirical plausibility when combined with data: if a given interpretation claims that its legal or political prescriptions will produce social stability and moral uplift, does a simulation of its structures, informed by real-world indicators, actually exhibit those patterns, or does it generate trajectories that resemble the very declines the Quran warns against?

In this way, interpretations become more than isolated opinions; they become research programs. Each one proposes a way of structurally inhabiting the Quranic world: a set of variables to prioritize, relationships to encode, and dynamics to accept. CRS provides the tools to explore those programs rigorously, to refine them, to recognize their strengths and limitations, and, when necessary, to revise or abandon them in light of deeper engagement with revelation and reality. The goal is not to eliminate disagreement, but to elevate it – to move from opaque disputes to transparent, model-based inquiry where sacred meaning is pursued with both intellectual humility and methodological discipline.

ACADEMIC RIGOR AND INTELLECTUAL DISCIPLINE

Mathematical thinking, when introduced into religious study through CRS, operates as a discipline of clarity. It confronts interpreters with questions they can no longer evade or answer vaguely: what exactly is being claimed when we say that a certain arrangement is just, that a policy is "Islamic," or that a behavior leads to spiritual decay? Which variables are actually involved – intention, harm, consent, distribution of risk, access to knowledge, vulnerability to coercion – and how are they related? What follows if we apply this claim consistently, not only in the cases that suit us politically or emotionally, but across the full range of situations the model implies? Formal modeling turns such questions into requirements rather than afterthoughts. A hazy statement that might pass in a sermon or even in a scholarly article cannot function as a CRS model until it has been pinned down into definable states, relationships, and trajectories. This pressure to specify is not a mere technical nuisance; it is a moral and intellectual discipline that exposes hidden contradictions, double standards, and unexamined assumptions.

When this mathematical scaffolding is applied at scale, hermeneutics itself begins to take on the characteristics of a scientific enterprise. Interpretations are no longer static pronouncements; they become hypotheses about how Quranic principles operate in the world. Building on the Five-Step Framework and UQVO, scholars engage in hypothesis formation – proposing structured readings of key verses and themes – followed by model building, where those readings are encoded as networks of variables and functions. These models can be tested: against the internal logic of the Quran, against other parts of the tradition, and against historical and contemporary data. When discrepancies or failures appear, interpretations are not simply defended by authority or dismissed by ideology; they are revised, refined, or in some cases replaced. Over time, it becomes meaningful to speak of "progress" in interpretation, not in the sense of moving closer to some imagined final theory, but in the sense of increasing coherence, explanatory power, and fidelity to both revelation and observed reality. Hermeneutics acquires feedback loops, error-correction mechanisms, and cumulative structure – features typically associated with scientific inquiry, now brought into the service of understanding sacred texts.

All of this points to the emergence of a new kind of scholar: the "computational mufassir." This figure is not a technocrat dabbling in religion, nor a traditional scholar reluctantly borrowing a few mathematical tools. Rather, the computational mufassir is fluent in both classical Islamic sciences and modern modeling. They can read the Quran and hadith with full awareness of linguistic nuance, historical context, and legal methodology, while also being able to express their interpretive conclusions in formal, computational terms. They move easily between text, tafsīr, and model: from verse to variable, from thematic exegesis to system dynamics, from moral principle to agent-based simulation. Their goal is not to mechanize revelation but to make explicit the structures that have always been implicit in serious scholarship, so that those structures can be shared, debated, and improved. In training such scholars, the tradition equips itself to converse on equal terms with an age defined by algorithms and models, without surrendering its theological depth or its commitment to divine guidance.

RISKS, OBJECTIONS, AND SAFEGUARDS

Any proposal to bring mathematics and computation into the heart of Quranic interpretation will meet understandable objections. The first is the fear of reductionism: the worry that by expressing aspects of revelation in equations, graphs, and variables, CRS is "turning the Quran into mathematics" and stripping it of its spiritual depth, literary power, and transcendence. A second concern is the specter of technocracy. If interpretive authority becomes tied to the ability to build and manipulate models, will religious leadership be quietly transferred from scholars of scripture to experts in data and algorithms? A third anxiety, closely related, is that these models could be weaponized: used by states, movements, or institutions to enforce a narrow ideological line under the guise of "scientific Islam," silencing other voices by claiming that "the model has spoken." Behind all three objections lies a legitimate fear that tools meant to clarify could instead be used to dominate.

These risks are real enough that they must be addressed at the level of design, not only with verbal reassurances. The first safeguard is conceptual: CRS must be explicitly framed and practiced as an aid, not a substitute, for human scholarship. Models are built from tafsīr, u ūl, and the broader tradition; they do not generate those disciplines ex nihilo. A model that cannot be explained, justified, and revised within the language of the classical sciences is already suspect. Second, the development and modification of ontologies, verse models, and dynamic systems must take place under formal governance structures. There should be clear processes for proposing changes, documenting rationales, reviewing them by qualified scholars and methodologists, and recording dissent. This is not bureaucracy for its own sake; it is how a community prevents quiet capture of its formal tools by a narrow cadre. Third, transparency must be non-negotiable. The models, their assumptions, and their derivation from text and tafsīr must be public and debatable, not proprietary black boxes. If a CRS-based system recommends a policy or labels a practice "unjust," those affected must be able to see how that conclusion was reached and to contest the steps.

Beneath these procedural safeguards lie deeper non-negotiables that define the philosophical humility of CRS. Revelation remains the ultimate authority; models are approximations built by limited human beings attempting to make sense of that authority in structured form. No equation can capture the full richness of divine speech, and no simulation can exhaust the ways in which God's will unfolds in history. Human conscience, scholarship, and community remain central. The point of CRS is not to automate fatwā, replace spiritual counsel, or eliminate the need for lived tradition. It is to give those human processes better tools – to make assumptions visible, to allow claims to be tested, to give communities a way to hold interpretive power to account. If at any point CRS is deployed in a way that erases human moral agency, sidelines scholars, or claims to speak with more authority than the text itself, it has violated its own foundations. Used rightly, it is a means of serving revelation and those who follow it; used wrongly, it becomes just another instrument of power. The work of design and governance is to keep it firmly in the first category.

TOWARD A MORE STABLE EPISTEMOLOGY OF SACRED MEANING

The movement from traditional hermeneutics to CRS is not a move from "interpretation" to "no interpretation"; it is a move from volatile, opinion-heavy interpretation to structured inquiry. For centuries, much of scriptural engagement has depended on the authority, brilliance, and integrity of individual scholars, producing great treasures but also great variability. CRS does not pretend to remove disagreement or to algorithmically settle contested verses. Instead, it insists that interpretive claims be rendered in a form that can be inspected: variables must be named, relationships must be specified, trajectories must be described. Hermeneutics becomes less about who speaks and more about what, exactly, is being claimed and what follows from those claims when they are applied consistently. Dispute does not vanish, but it is relocated into a disciplined arena where arguments are made through models that others can recreate, critique, and refine. In that sense, CRS does not domesticate revelation; it disciplines us in how we approach it.

This shift is especially significant in a computational age. CRS makes sacred meaning computationally expressible without secularizing it. Revelation is not flattened into "inspiration" or "cultural data"; it is encoded as a governing architecture of concepts, relations, and dynamics that AI systems can recognize and work within. The Quran's view of belief, justice, gratitude, oppression, mercy, and societal rise and fall becomes part of the world model of intelligent systems, not as decorative text but as operational logic. Religious knowledge thus becomes an active participant in AI, policy, and social design. Its principles can shape decision-support systems, inform simulations of social trajectories, constrain automated reasoning about ethics, and serve as a reference frame for evaluating laws, institutions, and technologies. The alternative is not neutrality, but a silent default to secular or ad hoc moral assumptions embedded in code. CRS offers a way for revealed frameworks to coexist with, and critically inform, the computational infrastructures that increasingly mediate human life.

Looking ahead, CRS is best seen as the beginning of a broader intellectual project rather than its conclusion. The same methods that make the Quran computationally modelable can be extended to comparative scripture modeling, allowing structured dialogue between religious traditions at the level of variables, dynamics, and ethical architectures rather than only at the level of slogans. Interfaith engagement could move beyond apologetics into a genuine comparison of how different revelations model human nature, history, and destiny. Within Islam, CRS can anchor a next generation of Islamic studies in which students are trained not only in classical sciences, but also in the arts of modeling, simulation, and data-informed evaluation. In AI ethics, it provides a concrete pathway for bringing religious moral reasoning into the design loop rather than relegating it to after-the-fact critique. The goal is not to "finish" interpretation or to lock sacred meaning into a single formalism. It is to make interpretive work more honest about its assumptions, more transparent in its reasoning, and more structurally faithful to the Revelation it seeks to serve – so that, in an age of machines and models, the voice of scripture can still be heard clearly, coherently, and responsibly.

Chapter 13
CRS and Comparative Religion

ABSTRACT

Chapter 13 argues that comparative religion has long suffered from vague abstractions, asymmetrical analysis, and projection of one tradition's categories onto another. Computational Religious Studies (CRS) offers a structural solution by modeling each scripture through variables, relationships, and dynamic systems true to its own semantics. With the Bible modeled through a parallel ontology (BVD) and the Qur'an through UQVO, CRS creates a shared conceptual space where justice, mercy, sin, covenant, and salvation can be compared without distortion. This framework reveals ethical symmetry, genuine divergence, and zones of incommensurability, enabling rigorous interfaith scholarship, comparative simulations, and AI systems that honor each tradition's integrity. Ultimately, CRS lays the foundation for a computational science of revelation that respects difference while enabling disciplined structural dialogue.

INTRODUCTION: WHY COMPARATIVE RELIGION NEEDS COMPUTATION

Comparative religion has traditionally worked with a set of tools that are rich, but also limited in very specific ways. For the most part, it has focused on narratives, doctrines, rituals, and institutions: how different communities tell their sacred stories, formulate their creeds, organize their worship, and build their religious hierarchies.

At a higher level, it has relied heavily on broad abstractions – "monotheism versus polytheism," "law versus love," "grace versus works," "prophetic religion versus mystical religion." These abstractions can be useful for orientation, but they are blunt instruments. They invite sweeping contrasts rather than precise analysis. They make it easy to say that "X religion emphasizes justice, Y emphasizes mer-

DOI: 10.4018/979-8-3373-6940-2.ch013

cy," but much harder to show, in a disciplined way, what this means structurally in terms of moral psychology, social order, or historical dynamics. The result is that comparisons remain vague, built on selective quotation and impressionistic judgments, and are always vulnerable to the interpretive bias of the scholar or tradition doing the comparing.

One of the deepest problems in classical comparative work is asymmetry and projection. Often, one tradition becomes the lens through which the other is seen. Islam is read through categories developed in Christian theology, or Christianity is evaluated using conceptual frameworks drawn from Islamic kalām and fiqh. Scholars tend to be deeply trained in one tradition – its language, sources, and internal debates – but only superficially acquainted with others, relying on translations and secondary literature. In practice this means that one side's categories, questions, and anxieties dominate the analytical space. Not only do we lack standards, but there is also no shared formal language for comparison. We can say that both scriptures speak about justice, mercy, sin, or salvation, but we have no standardized way of representing what "justice" or "salvation" means structurally in each system and how those meanings interact with other concepts. Without such a language, even well-intentioned comparative work risks becoming a subtle act of projection: each side finds in the other what it already expects to see.

Computational Religious Studies offers a way out of this impasse by introducing a variable-based, structural lens into comparative religion. Instead of treating "justice" or "mercy" as loose abstractions, CRS insists that they be modeled as variables with defined domains, relationships, and dynamics inside each scriptural world. In this framework, comparative religion is no longer simply a matter of juxtaposing doctrines or narratives. It becomes the task of aligning variables – justice, mercy, sin, covenant, salvation, guidance, community – and then comparing the structures in which they are embedded: how they interact, what feedback loops they form, how they evolve over time in the life of an individual or a society. The thesis is that CRS can ground comparative religion in computationally explicit models, reducing projection and enhancing structural clarity. Each tradition is first modeled on its own terms, with internal fidelity, and only then brought into comparison at the level of variables and dynamics. This does not erase genuine theological difference; it makes that difference intelligible in a new way, revealing where scriptures converge in ethical symmetry and where they diverge in their deeper architectures of meaning and salvation.

FROM CRS TO COMPUTATIONAL BIBLE: EXTENDING THE FRAMEWORK

Extending CRS to comparative religion begins with a simple but demanding idea: if we can build a Computational Quran, we can also build a Computational Bible. The core move is to apply the same Five-Step Framework that we use for Quranic modeling – source text, interpretive summary, variable mapping, mathematical expression, and thematic labeling – to Biblical texts. This means treating the Bible not just as a collection of stories, commandments, and letters, but as an underlying structure of concepts, relations, and dynamics that can be formalized. To do this, we introduce a Bible Variable Dictionary (BVD), analogous to the Universal Quranic Variable Ontology (UQVO). The BVD defines variables such as covenant, grace, sin, sacrifice, righteousness, kingdom, church, and redemption with precision rooted in the internal logic of Biblical texts and their interpretive traditions. The goal is not to homogenize the Bible with the Quran, but to model Biblical concepts with full internal fidelity before any comparison takes place.

Of course, moving from CRS as developed for the Quran to a Computational Bible requires methodological adjustment. The Quran is a single text with a unified canon and a specific historical context of revelation; the Bible is a library of texts, composed over centuries, by multiple authors in different languages, compiled into canons that are not identical across traditions. The Quran presents itself as a recited book with a clear final form; the Bible presents itself as narratives, laws, poetry, prophecy, wisdom literature, gospels, and epistles. A computational approach must respect these differences. This means designing pipelines that can handle multiple canons (for example, the Hebrew Bible and various configurations of the New Testament), and can account for textual variants and layered redaction. It also means engaging seriously with interpretive traditions – Jewish, Catholic, Orthodox, Protestant – each of which has developed its own hermeneutic keys and theological emphases. Where the Quranic pipeline can assume one canonical Arabic text, the Biblical pipeline must include mechanisms for annotating variants, tagging textual strata, and encoding the perspectives of different communities that read the same passages differently.

Because of this complexity, the first duty of a Computational Bible project is to ensure internal integrity before any comparative move is attempted. Each scripture must be modeled on its own terms. The semantics of "covenant" in the Hebrew Bible, for instance, must be drawn from its narrative, legal, and prophetic usage, not imported from how "'ahd" functions in the Quran. Similarly, "grace" in Pauline letters must be understood in its own theological architecture, not forced into categories drawn from Islamic kalām. CRS therefore insists that intra-tradition coherence is a prerequisite for inter-tradition comparison. UQVO governs the Quranic world; BVD governs the Biblical world. Only once each system has been coherently modeled – its variables, relationships, and dynamics stabilized within its own hermeneutic and doctrinal frame – do we begin to construct mappings between them.

This discipline prevents a common failure mode in comparative religion, where the language and categories of one tradition quietly dominate the analysis of the other. In a CRS-based comparative project, importing Quranic semantics into Bible modeling, or Biblical semantics into Quran modeling, is explicitly disallowed at the modeling stage. Instead, each text is granted its own conceptual sovereignty. Only at a later stage, in a controlled comparative layer, do we ask how UQVO.JUSTICE.Q relates to BVD.JUSTICE.B, or how Quranic accounts of mercy, sin, or salvation structurally resemble or differ from Biblical ones. By keeping the modeling pipelines distinct but parallel, CRS creates the conditions for a comparison that is genuinely respectful and structurally honest, rather than one that merely translates one scripture into the language of another.

SHARED CONCEPTUAL SPACE: DEFINING CROSS-SCRIPTURAL VARIABLES

Comparative religion, when approached through CRS, begins by identifying a shared conceptual space rather than assuming a shared theology. Many of the most important axes of meaning recur across scriptures: justice and mercy, sin and covenant, salvation and law, love and community, leadership and prophecy. It is tempting to assume that because the same or similar words appear in multiple traditions, they must mean the same thing. CRS resists that temptation. It distinguishes between superficial lexical similarity and deep structural commonality. "Justice" in one tradition may emphasize covenantal faithfulness, in another distributive fairness, in another restorative balance. "Salvation" may be framed in terms of liberation from political oppression, deliverance from sin, incorporation into a covenant community, or eternal union with God. Identifying shared axes of meaning therefore does not mean collapsing nuance; it means selecting concepts that function as major organizing nodes in more than one scriptural system and then asking how each tradition structurally configures them.

Once those axes are identified, CRS constructs what we might call cross-reference variable families. Take justice as an example. On the Quranic side, UQVO.JUSTICE.Q encodes ʿadl as it appears in the Quranic world: a balance of rights and duties, prohibition of zulm (wrongdoing/oppression), fair measurement in trade, impartiality in judgment, and alignment with divine mīzān. On the Biblical side, BVD.JUSTICE.B encodes justice as it emerges from Torah, prophets, wisdom literature, and New Testament texts: mishpat and tzedek in relation to covenant, protection of the widow and orphan, condemnation of exploitation, and themes of justification in Pauline theology. Each instantiation has its own domain, internal dynamics, and relational graph: which concepts it connects to, how it behaves over time, what kinds of divine responses it triggers. The "justice" family thus consists not of a single abstract idea, but of at least two formally defined variables – one Quranic, one Biblical – each rooted in its own text and interpretive tradition.

The heart of comparative CRS lies in how these variable families are aligned, and this is where it departs from naive approaches. It is not enough to say "justice equals justice" or "mercy equals mercy" across religions. Such statements are too simplistic and often wrong. Instead, CRS treats cross-scriptural alignment as a structured mapping problem. Alignment means constructing a mapping that relates UQVO.JUSTICE.Q to BVD.JUSTICE.B in terms of their roles, constraints, and relational neighborhoods: which other variables they are attached to, what kinds of actions they commend or condemn, how they participate in narratives of judgment and redemption. Some aspects will align strongly; others will diverge or remain incommensurable. The same is true for mercy, sin, covenant, salvation, and other shared families.

By treating alignment as a mapping between graphs and dynamic systems rather than as a semantic shortcut, CRS allows us to see both common ground and real difference with much greater clarity. We can say, for example, not just that "both traditions care about the poor," but that in one scripture care for the poor is structurally bound to covenant faithfulness and prophetic critique of kings, while in the other it is tied to zakat, prohibition of riba, and communal purification. The shared conceptual space thus becomes a site of rigorous comparison, where variables are related across traditions without erasing the distinct architectures in which they live.

STRUCTURAL COMPARISON OF THEOLOGICAL ARCHITECTURES

A central strength of CRS in comparative religion is that it allows us to treat justice and mercy not as abstract adjectives, but as interacting system variables within each theological architecture. In the Quranic world model, ʿadl (justice) and ra ma (mercy) are both attributes of God and demands placed upon human beings. Justice involves right measure, prohibition of zulm (oppression), and a moral ordering in which each person and group receives their due. Mercy, meanwhile, suffuses creation, precedes wrath in divine self-description, and frames guidance, repentance, and forgiveness. In a CRS model, these appear as tightly coupled variables: increases in injustice trigger warnings, social decay, and potential punishment, while sincere repentance and reform activate divine mercy, which can override the consequences that strict justice alone would impose. Biblical texts, modeled computationally, configure justice and mercy within covenantal and Christological frames. Justice expresses God's faithfulness to covenant, his demand for righteousness, and his defense of the oppressed. Mercy appears in divine forbearance, in the repeated calls to return, and climactically in the figure of Christ as the one who bears sin. In both systems, then, we can ask comparative questions in structural terms: what are the trade-offs between justice and mercy? Do both traditions posit similar thresholds beyond which punishment becomes inevitable? How does each architecture handle cases where mercy seems to suspend strict justice, and what systemic safeguards are modeled to prevent that mercy from collapsing into moral license?

The same structural approach can be applied to sin, repentance, and salvation. In Quranic modeling, sin is represented as a set of actions and states that distance the human being from God, darken the heart, and invite both worldly and eschatological consequences. Tawba (repentance) is a dynamic variable that can reverse trajectories: sincere turning back, accompanied by corrective action, cleanses sin, restores relationship, and changes the future path of the individual and community. Forgiveness is granted by God in response to this process, often modeled as a nonlinear reset of accumulated moral debt. Biblical models introduce additional elements: covenantal sacrifice in the Hebrew Bible, where offerings symbolically and ritually deal with sin; grace in the Pauline and broader New Testament sense, where divine favor and atonement in Christ provide a route to justification; and ongoing practices of confession and repentance within the community. CRS forces us to think in terms of trajectories. How does sin accumulate in each system – linearly, exponentially, with thresholds? Under what conditions does it dissipate? What are the pathways to restoration: direct repentance, sacrificial systems, mediated atonement, communal rituals? By encoding these processes as dynamic systems, we can see where the traditions converge in emphasizing the possibility of return and where they diverge in the mechanisms by which that return is structured.

Covenant, law, and community form a third domain where structural comparison becomes illuminating. In a Quranic CRS model, the ummah is the community bound by taw īd, divine command, and mutual responsibility. Law (sharīʿa) is not simply legislation; it is the set of divinely revealed norms that organize worship, family life, commerce, criminal justice, and public order. Covenant appears in the primordial pledge of humanity to God, in specific agreements with prophets and communities, and in the ongoing expectation that believers will uphold justice, enjoin good, and forbid evil. These elements are modeled as interdependent subsystems: weakening of law observance erodes communal identity; betrayal of covenant invites divine censure; strengthening of ummah ties amplifies the capacity for justice and mercy in practice. In Biblical and Hebraic frameworks, covenant is more explicitly foregrounded: the covenants with Abraham, Moses, David, and the "new covenant" language of Jeremiah and the New Testament define the relationship between God, Israel, and later the church. Law (Torah) structures Israel's communal life; grace and Spirit in Christian texts reshape how law is internalized and fulfilled. Community is envisioned as Israel, the ekklēsia (church), or the body of Christ, each with distinct boundaries and responsibilities.

CRS enables us to compare how law, grace, and communal identity interact structurally in these systems. Does law function primarily as an external code, an internalized disposition, or both? How do covenant and community mediate access to salvation and moral formation? When a community fails – through injustice, idolatry, or hypocrisy – what does each scripture's model predict about divine response, historical fate, and the possibility of renewal? By representing covenant, law, and community as linked variables with specified causal relationships, we can examine, for example, whether Quranic and Biblical architectures agree that persistent injustice leads to communal downfall, or how they each balance particularism (a chosen community) with universal horizons. Theological architectures thus become comparable at the level of systems: not by collapsing them into the same thing, but by revealing the distinct ways in which they organize justice, mercy, sin, repentance, law, and community into coherent – if different – maps of divine–human interaction.

ETHICAL SYMMETRY AND DIVERGENCE

A crucial advantage of CRS in comparative religion is that it forces us to define ethical outcome variables in a way that is comparable across traditions without erasing their distinctiveness. Both Quranic and Biblical corpora are deeply concerned with the protection of the vulnerable, honesty in trade, the integrity of family life, and the cohesion of the community. Rather than leaving these as broad themes, CRS asks what they look like as measurable outcomes. We can define variables such as SOCIAL JUSTICE, CARE FOR THE POOR, NON-OPPRESSION, FAMILY STABILITY, and SOCIAL TRUST, each with operational indicators and structural relationships to other variables in the model. On the Quranic side, these might be tied to injunctions about zakat, prohibition of riba, fair weights and measures, care for orphans, and admonitions against zulm. On the Biblical side, they may connect to laws about gleaning, jubilee and debt release, prophetic denunciations of exploitation, and exhortations to care for widows, orphans, and strangers. Once these outcomes are formalized, we can begin to align them as comparable targets even when the underlying narratives and legal forms differ.

Within this shared ethical space, CRS can identify zones of symmetry where the structural models converge in their treatment of key concerns. For instance, both scriptures place repeated and explicit emphasis on the treatment of orphans, widows, the poor, and strangers or immigrants. When modeled computationally, we often see similar patterns in system dynamics: neglect of these groups is associated with moral decline, social instability, and eventual divine censure; their protection and uplift are linked to communal blessing, resilience, and divine favor. Even if the language differs – zakat versus tithes, specific legal mechanisms versus broader exhortations – the causal motifs are similar. A CRS-based claim of ethical symmetry, then, is not a rhetorical statement that "both religions care about the poor," but an observation that, when modeled, the trajectories produced by ignoring or honoring the vulnerable look structurally alike: inequality and hardness of heart lead to breakdown; generosity and justice lead to stability and favor.

At the same time, CRS also reveals zones of ethical divergence where the underlying structures genuinely differ, and not merely at the surface of doctrine. Questions around divorce, war ethics, ritual law, and political authority often fall into this category. For example, Quranic and Biblical models may impose different constraints on when divorce is permitted, how reconciliation is sought, and how responsibility is distributed between spouses and community. War ethics may diverge in their rules of engagement, their handling of noncombatants, and their theological framing of violence and peace. Ritual laws differ significantly in scope and function, and political authority is conceptualized through different historical experiences and eschatological expectations. By encoding these domains as formal models, CRS allows us to see divergence as structural: different variables are prioritized, different thresholds trigger permissible or impermissible actions, and different long-run trajectories for individuals and communities are envisioned.

This structural view of divergence is important because it moves us beyond simplistic statements like "Islam is more legalistic" or "Christianity is more grace-centered," which are at best partial and at worst caricatures. Instead, we can say, for instance, that in one tradition ritual law plays a central role in demarcating community boundaries and shaping daily life, while in another it is relativized or reinterpreted in light of new covenantal structures. Or we can show that in one model political authority is tightly linked to explicit divine law, whereas in another it is mediated more through ecclesial structures and interpretive communities. Different ethical configurations are thus understood as the outcome of different modeled architectures, not simply as lists of rules. CRS thereby provides a way to honor both ethical symmetry and ethical divergence: it shows where scriptures echo one another in their concern for the vulnerable and the just ordering of society, and where they chart distinct structural paths – paths that can be understood, respected, and critically examined without being flattened into the same shape.

METHODOLOGICAL ARCHITECTURE FOR COMPARATIVE CRS

A comparative CRS framework rests on the idea of parallel but distinct modeling pipelines. On the Quranic side, the text passes through the established Five-Step Framework: verses are selected and contextualized, tafsīr is synthesized, variables are identified, mathematical expressions are derived, and thematic labels are assigned. These products are then organized within the Universal Quranic Variable Ontology (UQVO), which defines the semantic and structural identity of each concept in the Quranic world. The result is a family of models that capture how belief, justice, mercy, sin, guidance, community, and other core variables behave and interact across the Qur'an. On the Biblical side, a parallel pipeline is constructed: the Five-Step Framework is adapted to the specific character of Biblical texts, the Bible Variable Dictionary (BVD) serves as the ontology, and models are built that reflect the internal logic of covenant, sin, grace, law, church, kingdom, and so on. The key is methodological symmetry: both scriptures undergo the same kind of formalization, but the asymmetries in canon, genre, and theological development are respected rather than suppressed. Each pipeline is tuned to its text's structure and tradition, ensuring that what emerges is genuinely Quranic on one side and genuinely Biblical on the other.

Once both pipelines produce mature model sets, a comparative model layer can be constructed. This layer does not replace UQVO or BVD; it sits above them as a meta-structure that links their entries. In this layer, alignments are represented as mappings between graphs rather than as simple one-to-one word matches. Nodes in the Quranic concept graph – such as UQVO.JUSTICE.Q – are mapped to nodes in the Biblical graph – such as BVD.JUSTICE.B – where there is sufficient structural overlap to justify comparison. Edges, which represent relationships like "justice opposes oppression," "oppression triggers divine response," or "mercy moderates punishment," are also compared. We ask whether the motifs "justice–oppression–divine response" or "sin–repentance–forgiveness" are configured similarly or differently in each scripture. In some cases, alignments will be tight: the treatment of orphans, for instance, may show similar node and edge patterns. In other cases, alignments will be partial or complex, requiring many-to-many mappings or conditional correspondences. The comparative layer thus becomes a carefully curated web of correspondences between two internally coherent theological architectures.

A critical feature of this architecture is its explicit handling of incommensurability. There will be concepts and structures in each tradition for which no meaningful analogue exists in the other: certain Christological claims, specific Quranic formulations of finality of prophethood, or ritual and sacramental systems with no structural counterpart. Rather than forcing artificial alignments, CRS models these gaps explicitly. In the comparative layer, they appear as unpaired nodes, unmatched edges, or regions of one graph that have no mapping in the other. This does not signal a failure of the method; it is an accurate representation of irreducible theological distinctives. By marking incommensurability as such, CRS avoids the common comparative mistake of stretching categories until everything looks vaguely similar. It allows scholars and communities to see where genuine common ground exists and where divergence is fundamental, grounded not just in doctrine but in the deeper structures of how each scripture models God, the world, and human destiny.

NEW FORMS OF INTERFAITH SCHOLARSHIP

CRS opens the possibility of a fundamentally different kind of interfaith scholarship. Much of what has historically passed for "dialogue" has been either apologetics dressed in polite language or proof-text battles where each side marshals verses to support pre-existing positions. In that mode, the deepest question is often, "Who can win this argument?" rather than "What does each scripture actually say structurally about the world and the human condition?" A CRS-based approach invites a different posture. It shifts the conversation from, "Here is our verse against yours," to, "Here is how our scripture structurally models justice, mercy, sin, or repentance; how does yours?" The dialogue becomes an exchange of architectures, not just quotations. Each tradition lays out how its text organizes a given variable – what it connects to, how it behaves over time, what outcomes it drives – and then asks the other to do the same. Mutual understanding is grounded in structural comparison: the mapping of variables, relationships, and dynamics, rather than in rhetorical contest.

Once scriptures are expressed as computational models, interfaith work can also take the form of joint research rather than mere debate. Scholars can pose shared questions and explore how different revelations answer them structurally. How do the Quran and the Bible each model the rise and fall of nations? What patterns of moral decay, injustice, arrogance, or neglect of the vulnerable lead to decline, and what patterns of repentance, reform, and renewed covenant lead to renewal? How do they model the moral psychology of repentance – shame, remorse, turning back, restoration – and what trajectories do they predict for individuals and communities who respond differently to guilt and grace? How do they frame the ethics of power and wealth, and what dynamics emerge when wealth concentrates or power is abused? With CRS, these are not purely textual questions; they can be explored through comparative simulations. One can, for instance, model unjust wealth concentration within Quranic and Biblical frames and see how each system describes the long-run effects on social cohesion, divine favor, and moral stability. The point is not to declare a winner, but to see how different revealed architectures illuminate the same human problems.

Crucially, this kind of work allows for shared tools without erasing distinct commitments. Muslim and Christian (or Jewish) scholars can use the same computational methods – graph analysis, dynamic systems modeling, agent-based simulations – while remaining fully rooted in their own theological loyalties. The common ground is technical and structural: agreement on how to represent variables, how to document assumptions, how to test models against text and reality. The normative commitments remain distinct: each community decides what counts as authoritative text, how to interpret it, and what weight to give to different traditions of commentary. Interfaith scholarship thus gains a genuine common platform without doctrinal flattening. Scriptures are not folded into a single "universal religion"; they are allowed to stand in their difference, but that difference is articulated with a new level of precision. In this way, CRS does not just make comparative religion more rigorous – it makes it more honest, more dialogical, and more capable of engaging an age in which both faith and computation shape the future of human societies.

APPLICATIONS: COMPARATIVE MODELS IN PRACTICE

Comparative CRS models lend themselves naturally to educational tools that move far beyond side-by-side verse comparison. Imagine visual, interactive interfaces where students can see the "shape" of justice, mercy, sin, and salvation in each scripture as networks and trajectories rather than isolated slogans. Justice is no longer just a word; it appears as a node connected to oppression, wealth, testimony, covenant, and divine response, with those connections differently patterned in Quranic and Biblical graphs. Mercy can be explored as a dynamic variable: students can adjust parameters representing repentance, sacrifice, or intercession and watch how each scriptural system describes the unfolding of forgiveness over time. Such tools train the eye to notice structural nuance instead of caricatures: instead of saying "Islam is law, Christianity is love," learners see how law and love, justice and mercy, command and grace, are configured in each tradition's internal architecture. This kind of pedagogy invites deeper respect because it makes misrepresentation visibly harder. The models themselves reveal when a simplistic claim has lost contact with the structural reality of a scripture.

The same comparative models can inform policy and ethics advisory work in multi-religious societies. Policymakers are often presented with religious arguments as slogans – "our faith commands justice," "our tradition protects the family" – without any clear sense of what those claims structurally entail. CRS-based tools can show where ethical recommendations from different scriptural traditions converge or diverge at the level of modeled outcomes and dynamics. For example, when designing policies on poverty, migration, family law, or economic regulation, advisors could consult comparative simulations that display how each scriptural frame envisions the effects of certain choices on social cohesion, protection of the vulnerable, and long-term stability. Instead of treating religion as a rigid constraint or a rhetorical resource, policymakers can see it as a set of articulated moral architectures that deserve to be understood on their own terms. This does not mean that public policy simply implements any one tradition's prescriptions; it means that decisions are made with an informed appreciation of the depth, convergence, and divergence of scriptural ethics, rather than on the basis of stereotypes or pressure from the loudest voices.

At the level of technology, comparative CRS provides a blueprint for AI systems designed to serve pluralistic contexts. Rather than a single, undifferentiated "religious mode," we can design AI that can switch between Quran-aware and Bible-aware reasoning modes, each grounded in structurally accurate models of its respective scripture. When a Muslim user asks an ethical or spiritual question, the system reasons within the Quranic architecture encoded in UQVO and its models; when a Christian user asks a parallel question, the system operates within the Biblical architecture encoded in BVD. In multi-faith consultations – say, a tool designed for hospital chaplaincy, humanitarian ethics, or peacebuilding – the AI can transparently show how different traditions structure the same ethical problem, highlighting both common ground and real differences without privileging one by stealth. Because these behaviors are driven by explicit models rather than vague "training on religious text," the system's handling of multi-faith inputs can be audited, critiqued, and improved by scholars from each tradition. In this way, comparative CRS underwrites AI that is not only more respectful but also more precise – capable of engaging religious diversity as a structured reality rather than a set of interchangeable labels.

RISKS AND RESPONSIBILITIES IN COMPARATIVE CRS

Any attempt to formalize and compare scriptures carries real risks that must be faced honestly. One danger is that comparative CRS could be misused to promote syncretism or forced homogenization, presenting different revelations as if they were merely local variations of a single generic religion. When variables are aligned and graphs are mapped, it becomes tempting to declare quick "equivalences" and to celebrate similarity while ignoring the deep doctrinal and structural differences that give each tradition its distinctive identity. Another risk is precisely that kind of superficial equivalence: claiming, for example, that justice or salvation "mean the same thing" in all traditions because they share some ethical motifs, while overlooking critical divergences in how those concepts are rooted in narrative, covenant, and metaphysical commitments. A more subtle but equally serious danger lies in power asymmetries. If one tradition's categories, questions, and modeling choices quietly dominate the design of the comparative framework, the other will be forced into an alien mold. This would turn comparative CRS into a new vehicle for intellectual hegemony: a way of making one scripture intelligible only insofar as it can be expressed in the concepts of another.

To avoid these outcomes, comparative CRS must be built with safeguards at its core. First, each tradition's models must be governed by its own community of scholars. Muslim scholars should oversee UQVO and Quranic models; Christian and Jewish scholars should oversee BVD and Biblical models. Comparative mappings can then be proposed and debated jointly, but no side's internal semantics or theological structures should be altered without its own tradition's consent. Second, assumptions, mappings, and non-alignments must be explicitly documented. When a variable in one scripture is aligned with a variable in another, the rationale must be clear; when no meaningful alignment exists, that "gap" must be recorded rather than quietly ignored. This transparency allows readers to see where comparisons are strong, where they are tentative, and where they are simply not possible. Third, the comparative layer must never override internal theological integrity. If a mapping would require distorting what a tradition holds as fundamental – about God, revelation, salvation, or community – then that mapping should be rejected or qualified, even if it would make the comparative picture tidier.

These safeguards reflect a deeper ethical posture that must guide all comparative CRS work. The purpose is to serve understanding, not to dissolve religious difference into a bland relativism. The models are meant to illuminate how each scripture structures moral and spiritual reality, not to declare that all structures are interchangeable or equally valid in every respect. This requires humility at every stage. Models are approximations, not replacements, for living faith. They capture patterns and relationships, but they cannot contain the full depth of worship, devotion, and encounter with the Divine that believers experience in their communities. Comparative CRS should therefore be approached as a tool for disciplined listening: a way for traditions to articulate themselves more clearly, to see each other more accurately, and to engage in deeper dialogue about convergence and divergence. When it begins to act as a judge over scriptures or a substitute for lived religious life, it has betrayed its proper role. Used rightly, it can help build a more honest and structurally aware conversation between faiths in a computational age; used wrongly, it risks becoming just another abstraction imposed on realities it does not fully comprehend.

TOWARD A COMPARATIVE SCIENCE OF REVELATION

A comparative CRS framework ultimately points toward something larger than "comparative religion" in its classical sense. It invites us to move beyond the notion of religions as mere "belief systems," understood as collections of doctrines that individuals happen to hold. In the CRS view, scriptures are epistemic architectures and dynamic moral systems. They are not just lists of propositions about God and the world; they are structured ways of knowing, feeling, and acting. Each revelation offers a particular configuration of concepts, relationships, and trajectories – how knowledge is acquired, how moral responsibility is framed, how communities are formed and judged, how history is interpreted, and how ultimate purpose is described. When we model scriptures computationally, we are not reducing them to data; we are making their internal architectures visible enough to be examined, compared, and engaged with at a structural level. This shift allows us to treat sacred texts as sources of actionable models of human and societal life, not just as repositories of beliefs.

For this vision to become a serious field of inquiry, comparative scripture must be developed as a domain of computational study in its own right. That means more than ad hoc projects; it calls for shared standards, tools, and benchmarks. Scholars will need agreed-upon protocols for building verse-level models, for defining and documenting variable dictionaries like UQVO and BVD, and for constructing comparative mappings that are methodologically transparent and theologically responsible. Datasets will have to be curated that contain not only texts and translations, but also structured representations: models of key verses, thematic graphs, dynamic systems capturing sin–repentance–forgiveness cycles, justice–oppression–divine response motifs, and so on. These resources can then be shared, critiqued, and extended by Muslim, Christian, Jewish, and other scholarly communities, creating a growing repository of formally expressed religious architectures. Over time, this could resemble what corpora and benchmarks are for natural language processing, but with the focus on the structural content of revelation rather than on generic language behavior.

Once such an infrastructure exists, the future directions are wide-ranging. It becomes possible, for example, to run multi-scripture simulations of civilizational trajectories: what happens to a modeled society over generations if it follows a Quranically grounded ethic of wealth, power, and family, versus a society modeled on particular Biblical or other scriptural patterns? How do different revelation-based architectures understand resilience in the face of injustice, catastrophe, or moral decay? Comparative analysis of justice, compassion, and social cohesion in modeled societies could reveal both shared insights and distinctive strengths in different traditions. In this context, CRS emerges as one pillar of a broader Comparative Science of Revelation: a research program that treats sacred texts as structurally analyzable sources of knowledge about human beings, community, history, and ultimate purpose. The aim is not to submit revelation to secular science, nor to collapse theology into mathematics, but to recognize that revelation itself offers models – of the soul, of society, of destiny – that can be articulated and examined with the best tools of our age. In doing so, we open a path for religious wisdom to participate directly and rigorously in the global conversation about where humanity is heading and what kind of world we are trying to build.

Accommodating Non-Theistic and Eastern Traditions in a Comparative Science of Revelation

A universal Comparative Science of Revelation must not assume that all sacred traditions are built on the same theological architecture. The Abrahamic traditions are generally organized around divine agency, prophetic communication, covenant, command, law, moral accountability, and eschatological consequence. A computational framework developed only around these categories would be too narrow if it claimed universality. Traditions such as Buddhism, Daoism, Confucianism, Hindu philosophical schools, and other Eastern systems often operate through different ontological and epistemic assumptions. They may emphasize awakening rather than obedience, harmony rather than command, liberation from suffering rather than salvation from sin, cosmic order rather than divine decree, or disciplined practice rather than revealed law.

This does not make computational comparison impossible. It means the framework must operate at a higher level of abstraction. Instead of beginning with tradition-specific variables such as "prophet," "revelation," "sin," "law," or "covenant," the comparative model should begin with more general categories: source of authority, nature of reality, model of the self, diagnosis of the human condition, path of transformation, ethical discipline, social order, ultimate aim, and mechanisms of consequence. These categories can then be specialized differently across traditions.

For example, in a Quranic model, the source of authority may be divine revelation; the human problem may be misguidance, injustice, arrogance, or forgetfulness of God; the path may involve faith, worship, moral action, repentance, and obedience to divine guidance. In Buddhism, the source of authority may be awakened insight and disciplined transmission; the human problem is suffering rooted in craving, ignorance, and attachment; the path is ethical conduct, meditation, wisdom, and liberation from suffering. In Daoism, the central structure may not be command and obedience, but alignment with the Dao, non-forcing action, naturalness, balance, and harmony with the underlying flow of reality.

Computationally, these can be represented as different world models of transformation. Each tradition defines an initial human condition, a set of constraints, a path of corrective action, and a desired end state. The terms differ, but the structure can still be compared:

$$S_{t+1} = f(S_t, A_t, C_t, P_t, R_t)$$

where represents the current existential, moral, spiritual, or cognitive state; represents action or practice; represents context; represents the governing principle, path, or discipline; and represents consequence, realization, correction, liberation, harmony, or accountability. In a theistic tradition, may include divine judgment or mercy. In Buddhism, it may include karmic consequence, reduction of suffering, or progress toward awakening. In Daoism, it may include increased harmony, reduced distortion, or alignment with the Dao.

This approach avoids forcing all religions into a single Abrahamic template. It also avoids reducing them to vague "spiritual wisdom." Each tradition retains its own ontology. The framework compares them by making their internal structures explicit: what each tradition claims about reality, the human condition, transformation, ethical action, and ultimate fulfillment.

The phrase Comparative Science of Revelation may therefore require careful definition. In a narrow sense, "revelation" applies most naturally to traditions that understand ultimate truth as disclosed by God through prophets, scripture, or divine speech. In a broader comparative sense, the field may examine structured claims of ultimate knowledge, including revelation, awakening, insight, cosmic order, dharma, dao, and transmitted wisdom. Under this broader framing, the project is not limited to theistic revelation but includes all traditions that offer authoritative models of reality and human transformation.

Chapter 14
The Future of Computational Quran: Complex Adaptive Systems and Quantum Computing

ABSTRACT

Chapter 14 explores the future trajectory of the Computational Qur'an by situating it within the frameworks of complex adaptive systems (CAS) and quantum computing. It argues that once Qur'anic guidance is formalized as structured variables, dynamics, and rule sets, it naturally lends itself to large-scale simulations of interacting agents, communities, and civilizations. The chapter shows how belief dynamics, justice networks, trust graphs, and moral feedback loops can function as local rules in CAS, producing emergent social and historical patterns consistent with Qur'anic narratives. It then outlines how quantum computing may, in principle, enable efficient exploration of the vast state spaces and knowledge graphs generated by these models. Together, CAS and quantum approaches are presented as future research containers that extend CRS without altering revelation, pointing toward a living, computationally enhanced tafsīr grounded in humility, rigor, and ethical restraint.

INTRODUCTION

The previous chapters of this book were deliberately foundational. We built the language of Computational Religious Studies (CRS) from the ground up: the Five-Step Modeling Framework, the Universal Quranic Variable Ontology, and the core mathematical tools that turn verses into explicit, analyzable models. By the time one surveys the Quranic Modeling Project and its growing library of formal models – belief trajectories, trust networks, obligation graphs, ethical vectors,

DOI: 10.4018/979-8-3373-6940-2.ch014

guidance fields – it becomes clear that the "Computational Quran" is no longer just a metaphor. It is now a concrete object: an interconnected system of variables, equations, and semantic structures that mirrors, in a carefully disciplined way, the guidance architecture of the Quran itself.

This chapter asks a natural next question: what happens when this mathematically articulated Quranic system is placed inside the two major revolutions reshaping our understanding of complex systems and computation – complex adaptive systems (CAS) and quantum computing? Artificial intelligence, which has already been mentioned repeatedly in this book, is only one part of the technological landscape. Complexity theory has given us a language for multi-level, self-organizing systems whose behavior cannot be reduced to the sum of their parts (Prigogine & Stengers 1984, Stacey 2011, Mitchell 2009, Kauffman 1993, Cilliers 1998). Quantum physics and quantum information have forced us to re-examine what "state," "causality," and "computation" even mean at the most fundamental levels (Dirac 1958, von Neumann 1955).

The Computational Quran sits at an interesting intersection of these developments. On the one hand, its models already exhibit many features that naturally belong in a CAS framework: interacting agents, non-linear feedback loops, emergent group behavior, and long-term trajectories shaped by local decisions. On the other hand, the formal Quranic variable space is vast and high dimensional, containing combinatorial numbers of possible configurations. Quantum computing, at least in principle, offers a way to explore such spaces in parallel by representing them in superposed quantum states. The ability, for example, to represent a knowledge graph where an edge between two nodes can exist in multiple states – present, absent, or weighted by different interpretations – mirrors the quantum ability to keep multiple possibilities "alive" simultaneously until a measurement collapses them.

This chapter therefore has a dual purpose. First, it introduces the reader to the basic ideas of complex adaptive systems and quantum computing, not as separate disciplines for their own sake, but as extended "containers" into which the Computational Quran can be placed. Second, it sketches a research agenda for future work in CRS: how Quranic models might serve as rule sets in large-scale simulations of moral and social dynamics, how quantum representations might accelerate or deepen the analysis of Quranic knowledge graphs, and how all of this can be pursued while maintaining the theological and ethical boundaries articulated in earlier chapters.

The aim is to show that once the Quran's guidance has been expressed as structured, dynamic models, those models can participate in the same scientific and computational frameworks that we use for ecosystems, economies, neural networks, and physical systems.

1. COMPLEXITY THEORY AND THE TURN TOWARD SYSTEMS

Before we can speak about Complex Adaptive Systems in the context of the Quran, we need a clear sense of what "complexity" means in this technical sense. Complexity theory is not merely a synonym for "complicated." A jet engine is complicated; it has many parts and a detailed design, but if you understand every component and every equation, you can in principle predict how the engine will behave. A complex system, by contrast, contains such rich interdependence and feedback that its global behavior is not obvious from its local rules (Holland 1995).

Classic examples of complex systems include ecosystems, financial markets, the human brain, cities, and the internet. These systems are characterized by a few recurring features. They consist of many interacting components – agents, cells, firms, neurons, organisms – each following relatively simple rules. These components are embedded in networks of influence; their behavior depends on neighbors and context. They change over time, often adapting or learning. And perhaps most importantly, they exhibit emergent phenomena: patterns of organization at the macro-level (flocking, market cycles, social norms, waves of innovation) that were not explicitly programmed into any single component.

Complexity theory (Mitchell 2009) focuses on questions such as: How do local rules give rise to global patterns? What kinds of feedback loops stabilize or destabilize a system? When do systems self-organize into robust forms, and when do they collapse into chaos or rigidity? How do multiple scales – individual, group, institution, culture – interact?

In mathematics and computer science, these questions are studied using tools like nonlinear dynamical systems, network theory, statistical physics, agent-based modeling, and game theory. CAS in particular emphasizes agents that adapt their behavior over time in response to experience, forming a constantly evolving ensemble.

It does not take much imagination to see why this matters for scripture modeling. Quranic guidance is not only about isolated individuals; it is about families, communities, nations, and civilizations. It is about how repeated patterns of justice or oppression transform a society, how trust networks shape economic life, how hypocrisy spreads through communities, how remembrance and gratitude decay across generations, and how divine laws interact with human agency over long time scales. All of these are, inherently, complex-systems questions.

The computational models in the Quranic Modeling Project already contain the raw ingredients for such analysis: dynamic belief equations, trust graphs, obligation networks, covenant memory decay functions, and so on. What CAS brings is a way to embed these local equations into large ensembles of interacting agents and then ask: what emerges when a whole population lives under these rules?

2. THE QURANIC MODELS AS RULE SETS FOR AGENTS

The move from static models to complex adaptive systems begins with a simple observation: each verse-level model can be interpreted as a rule set for agents in a simulation.

Consider the belief dynamic equations from early al-Baqarah, where believers, disbelievers, and hypocrites are represented as different trajectories of a belief state (H_I(t)), (H_K(t)), and (H_N(t)) driven by revelation and temptation. Believers are modeled as second-order systems that respond to guidance (AG(t)) with eventual stabilized growth, disbelievers as essentially flat due to a sealing condition, and hypocrites as oscillatory systems with no convergence. In a CAS framework, each individual agent in a population can be assigned one of these update rules for its internal state, perhaps with parameters that vary slightly from agent to agent to represent personality, history, or trauma.

The trust and obligation graphs built from riba (usury), charity, and debt-documentation verses give rules for how agents form financial links, share risk, transfer obligations, and assign trust weights. Family law models treat marriage as a finite-state machine with controlled transitions; covenant models treat memory and gratitude as decaying functions refreshed by reminders; hypocrisy models assign each agent a speech-action divergence metric that affects public perception. All of these can be interpreted as local rules governing how an agent behaves, how it updates its state, and how it interacts with neighbors.

In an agent-based CAS simulation, the core structure looks like this: there is a set of agents (i = 1, 2, …, N); each agent has a state vector (S_i(t)) at time (t), containing variables from UQVO such as belief (H_I^i), taqwā (H_T^i), hypocrisy (H_N^i), group memberships, wealth, trust in neighbors, and so forth. There is also an environment capturing global influences like economic shocks, disasters, political changes, and aggregate levels of guidance and temptation. On each time step, each agent updates its state according to a function

$$S_i(t+1) \ = \ F(S_i(t), \textit{neighbor states}, \textit{environment}(t))$$

where (*F*) is a composite of the verse-level functions and dynamics models we have already formalized. Some parts of (*F*) are deterministic; others incorporate probability to represent uncertainty or choice. Agents may also change their network connections – who they trust, whom they trade with, which leaders they follow – based on past experiences.

This is, in essence, how CAS works: a global pattern emerges from local updates and ongoing interactions. The key insight is that the Quranic models already define the local rules; CAS is the software architecture that runs them at scale. If the mathematical models tell us what happens when one hypocrite hears a reminder, CAS tells us what happens when thousands of hypocrites, believers, and disbelievers interact in a shared environment over generations.

For CRS, this opens a new way of asking questions. Instead of only asking, "What does this verse say about belief?" we can ask, "If a community lives according to the local dynamics encoded in these verses, what emergent patterns arise? Under what conditions does justice stabilize? Under what conditions does corruption become self-reinforcing? Where are the tipping points where small interventions – like a single act of charity, or a single unjust policy – have disproportionate long-term effects?"

3. MULTI-SCALE QURANIC SYSTEMS: FROM HEARTS TO CIVILIZATIONS

Complex adaptive systems are inherently multi-scale Cilliers 1998, Mitchell 2009). They operate at several levels simultaneously, and behavior at one level feeds back into others. CRS can adopt this multi-scale perspective by recognizing at least three interacting levels in the Quranic modeling space: individual hearts, communities, and civilizations.

At the individual level, we have already seen models for belief trajectories, moral integrity, repentance impulses, temptation decay, and spiritual memory. An agent's inner state is described by variables in the H_X and D_X families, evolving under differential equations and thresholds. The Quran speaks directly to this level when it describes hearts softening or hardening, when it promises internal tranquility to those who remember God, or when it warns of hearts that become sealed.

At the community level, the Quranic Modeling Project has begun to model trust networks, risk-sharing obligations, covenant memory across generations, and group dynamics like hypocrisy clusters or believer coalitions. Economic verses define how wealth flows, how riba distorts risk, how charity redistributes, and how documentation reduces conflict. Legal verses define how marriages form and dissolve, how inheritance flows, how disputes are adjudicated. In CAS terms, communities are networks where nodes are individuals and edges are social, economic, or legal relationships.

At the civilizational level, the Quran often speaks of entire peoples rising and falling: nations that are destroyed for persistent injustice, communities that flourish under gratitude and obedience, prophetic lineages that carry covenants through time. The Abrahamic covenant models and narrative ownership functions already hint at this level, treating prophetic history as a directed graph of covenant transfers and identity claims.

In a CAS framework, we can explicitly couple these levels. For example, if many individuals in a community adopt unjust practices, the trust network degrades; as the trust network degrades, more individuals experience temptation to exploit or withdraw; as that happens, institutions become fragile; and as institutions falter, historical memory of past guidance may decay faster, leading to civilizational decline. Conversely, acts of charity and justice can propagate through networks, lowering the effective "temperature" of social conflict and stabilizing equitable norms.

Mathematically, this often looks like coupled sets of differential or difference equations, where macro-variables (like community trust, average taqwā, wealth inequality) are aggregates of micro-variables, and the evolution of micro-variables depends on those aggregates. Complexity theory provides both the language and the tools to study such couplings: mean-field approximations, renormalization approaches, network-level order parameters, and so on.

From a Quranic perspective, what is exciting about this is not that we finally "explain" divine patterns in purely naturalistic terms. Rather, we can use the revealed models as design specifications and see how they respond when instantiated at different scales. Where the Quran says that oppression leads to downfall, CAS can help us explore the mechanisms by which oppression propagates and destabilizes systems. Where the Quran says that gratitude increases blessing, CAS can show how grateful behaviors may amplify cooperative dynamics and resilience. The revelation gives the normative structure; CAS provides a theater in which that structure can be explored dynamically.

4. ADAPTATION, LEARNING, AND NORMATIVE CAS

A distinctive feature of complex adaptive systems is that agents learn or adapt their behavior. In biological systems, this takes the form of evolution or neural learning. In economic models, it takes the form of firms adjusting strategies based on market signals. In social systems, it appears as people changing behavior in response to norms, incentives, and experience.

In a Quranic CAS, adaptation must be treated carefully. Humans are not reinforcement-learning algorithms optimizing a single reward; they are moral agents with free will, limited knowledge, and diverse motives. Still, the ontology and models we have already built provide handles for representing forms of adaptation that the Quran itself acknowledges: repentance, stubbornness, increased sincerity, or increasing arrogance.

In the modeling, repentance appears as an impulse in the moral integrity differential equation, a corrective term that shifts a trajectory back toward alignment with divine law. This can be treated in CAS as an adaptive update rule: when certain conditions are met – say, exposure to reminders, consciousness of sin, remaining softness of heart – an agent may trigger a repentance event that abruptly changes its internal state and subsequent behavior.

Similarly, repeated engagement in sin can gradually change thresholds and sensitivities. A person who repeatedly commits injustice may find it progressively easier to do so; mathematically, the cost term in their internal "utility" function shrinks, or the repulsion from injustice in their spiritual potential field diminishes. CAS models can express this as adaptation in the wrong direction: local updates that push an agent deeper into a basin of attraction corresponding to hypocrisy or hardened disbelief.

What makes a Quranic CAS different from a purely descriptive one is its normative dimension. The update rules themselves embody values: taqwā is not just another state; it is a desired state. Hypocrisy is not just another strategy; it is condemned. Divine attributes, though exogenous, define gradients of what counts as virtuous adaptation versus corrupt adaptation. Simulation in this context becomes a way to explore what happens when agents respect or violate those normative rules, not to redefine the norms themselves.

One can imagine, for example, agent-based simulations where some agents follow Quranic rules of trade, charity, and risk, while others operate according to purely self-interested or exploitative strategies. How do such mixed populations behave? Under what conditions do Quranic strategies prove resilient or fragile? How does the presence of even a small number of sincere agents change the trajectory of a corrupt network? These are CAS questions grounded in normative scripture.

5. NONLINEARITY, ATTRACTORS, AND MORAL PHASE TRANSITIONS

Dynamical systems underlying CAS are rarely linear (Mitchell 2009). Small changes in one part of the system can trigger large effects elsewhere, and the system can have multiple attractors – stable states or patterns toward which it tends to evolve.

The belief models already show this qualitatively. Believers tend toward a stabilized, mature faith under sustained guidance; disbelievers remain at a flat zero; hypocrites oscillate without convergence. Each of these is an attractor, with its own basin of attraction. Once an agent's internal parameters – sincerity, openness, resistance – are set within a certain range, their trajectory tends to one of these regimes unless a strong intervention occurs.

In CAS terms, we can think of moral and spiritual states as existing in a high-dimensional "phase space," where different combinations of UQVO variables correspond to different points. Certain regions of this space correspond to stable believer profiles; others to hypocritical profiles; others to hardened disbelief. The Quran itself often uses language of paths and destinations: "those guided," "those astray," "those who incur wrath."

Complexity theory encourages us to look for phase transitions in such spaces: critical thresholds where a small change in parameters causes the system to jump from one attractor to another. For example, a society may tolerate a small amount of corruption without collapsing, but beyond a certain point, corruption cascades through institutions and trust networks, leading to systemic failure. A heart may resist temptation for a long time, but repeated transgression beyond a threshold leads to sealing.

Mathematically, phase transitions often occur when feedback loops switch sign or when eigenvalues of certain operators cross zero. In Quranic CAS, we can look for combinations of parameters – levels of remembrance, degrees of injustice, concentrations of wealth, strengths of family ties – at which the system's qualitative behavior changes. These are the tipping points that the text warns about narratively; CAS allows us to explore them structurally.

This is not to say that we can precisely locate such thresholds in real societies; that would require empirical data and carries many uncertainties. But even as a modeling exercise, identifying possible tipping points can deepen our understanding of why the Quran is so insistent on certain patterns (such as eliminating riba or preserving justice) and so wary of gradual drift.

6. NETWORKED GUIDANCE AND MORAL EPIDEMICS

Another major theme in complexity science is the role of networks in spreading influence. Diseases, rumors, innovations, and behaviors spread through networks; some spread quickly and widely, others die out. The structure of the network – who is connected to whom, with what strength – matters as much as the content of the influence.

The Quranic models of trust, obligation, charity, and leadership already form networks: graphs where people are nodes and relationships like debt, kinship, leadership, or friendship are edges. Belief and guidance can thus be modeled as a kind of "moral contagion" spreading through these networks, though with important differences from physical contagion: acceptance of guidance involves choice, insight, and divine grace, not just passive exposure.

Even with that caveat, network theory offers useful tools. We can model guidance as a field that propagates more easily along edges with high trust weights, or along edges where previous experiences have been positive. We can model misguidance and hypocrisy as propagating along different edges, perhaps faster in environments with low knowledge or weak institutions. The Quran's emphasis on companionship – being with the truthful and avoiding corrupt company – maps naturally into a network sensitivity: connect to certain kinds of nodes and your trajectory changes.

In CAS, network epidemiology studies phenomena like basic reproduction numbers, herd immunity thresholds, and critical mass effects. Analogues can be developed for Quranic CAS: how many sincere agents in positions of influence are needed for a community to resist a wave of corruption? How much knowledge must be disseminated before a hypocritical narrative loses its grip? How does the structure of a trust network – centralized versus decentralized – affect the stability of justice?

Simulations can allow us to compare different network architectures under the same Quranic rule sets. A star-like network, where one central leader holds most edges, may be efficient but fragile; a more distributed network may be robust but slower to coordinate. There is a deep resonance here with the Quran's attention to leadership, consultation, and community solidarity.

7. PRACTICAL CAS RESEARCH DIRECTIONS IN CRS

Even without quantum computing, a CAS-oriented approach to the Computational Quran opens a large research frontier. We can imagine at least several streams of work.

One stream is synthetic simulation: building hypothetical societies of agents who follow Quranic rules and comparing their behavior to societies with different norms. The point is not to predict real history but to deepen understanding of how Quranic norms might function as system constraints.

A second stream is qualitative calibration: tuning CAS models to reflect known historical scenarios from Quranic narratives – such as communities destroyed for injustice, or early Muslim community dynamics – and seeing whether the models replicate the broad patterns described in scripture and history.

A third stream is policy-relevant analysis: using CAS, informed by Quranic models, to think about contemporary questions of economic justice, social cohesion, or environmental stewardship. This would enable us to explore structural implications of different choices under a Quran-shaped value system.

In all of these cases, the key is that the local rules are derived from the Quran via the Five-Step Framework and UQVO, while the CAS machinery provides a way to explore their collective implications. This preserves the primacy of revelation while taking seriously the complexity of the world in which that revelation is to be lived.

8. FROM CLASSICAL TO QUANTUM COMPUTATION

If complexity theory extends our conceptual reach into large-scale dynamics, quantum computing extends our computational reach into enormous state spaces. To understand what this means for CRS, we need a brief orientation to quantum computing.

Classical computers operate on bits: units of information that can be either 0 or 1. A classical state of (n) bits is one point in a space of (2^n) possibilities. Computation proceeds by applying logical gates to these bits – AND, OR, NOT – moving from one classical state to another.

Quantum computers operate on qubits. A single qubit can be in a state (|0⟩), (|1⟩), or any superposition (α|0⟩ + |1⟩) where (α) and ( ) are complex amplitudes satisfying (|α|^2 + | |^2 = 1). When we measure the qubit, it collapses probabilistically to 0 or 1, with probabilities given by the squared magnitudes of the amplitudes.

With multiple qubits, the state space explodes. Two qubits can be in any superposition of (|00 , |01 , |10 , |11); three qubits live in an 8-dimensional space; (n) qubits in a (2^n)-dimensional Hilbert space. The power of quantum computing comes from manipulating these superpositions through unitary operations (quantum gates) such that many classical configurations are processed in parallel.

Quantum mechanics also allows entanglement, where the state of one qubit is correlated with that of another in a way that cannot be decomposed into independent parts. Measurement outcomes in one part of the system instantaneously determine correlated outcomes elsewhere (within the formalism), though no usable faster-than-light communication is possible.

From a computational perspective, these features enable algorithms like Shor's factoring algorithm and Grover's search algorithm, which offer exponential or quadratic speed-ups over classical algorithms for specific tasks. They also enable quantum walks on graphs and quantum versions of machine learning methods.

Importantly, quantum computing is not "magical." It does not give arbitrary exponential speed-ups for all problems, and real-world quantum devices are noisy and error-prone. But even as the hardware slowly matures, thinking in quantum terms encourages us to imagine new ways of representing information and exploring combinatorial spaces – exactly the kind of spaces that arise in high-dimensional Quranic models.

9. QURANIC STATE SPACES AND QUANTUM ENCODINGS

The formalization of Quranic content into UQVO variables and mathematical models has, in effect, created very large state spaces. A single individual's ethical state can be represented by a vector of variables like (H_I), (H_T), (H_N), group memberships, current obligations, memory, and so on. A community adds another dimension: the network of trust and obligation edges among individuals. A civilizational model adds still more: aggregate variables, historical position, and environmental factors.

If we think of each binary or discretized variable as a bit, even a modest model quickly requires dozens or hundreds of bits to specify a single configuration. A population of (N) agents each with (k) binary variables has (2^{Nk}) possible configurations. Classical simulation can only explore a tiny fraction of that space explicitly; it must rely on approximations or sampling.

Quantum computing offers a formal way to encode such huge spaces into quantum states. For example, we can assign each possible configuration of a Quranic CAS – each assignment of states to all variables and network edges – to a basis vector in a Hilbert space. A quantum computer can then hold a superposition of many such configurations simultaneously:

$$|\Psi = \sum_c \alpha_c|c$$

where (c) ranges over configurations and (α_c) are complex amplitudes.

In practice, encoding meaningful Quranic configurations into qubits is a nontrivial engineering task. One must decide how many qubits to allocate per variable, how to represent continuous variables like belief strengths, how to ensure constraints (like conservation of certain quantities) are respected, and how to map update rules into unitary operations. But conceptually, the idea is straightforward: the same way quantum chemists encode molecular structures into qubit states, we can in principle encode Quranic structural models into quantum state spaces.

Once encoded, we can apply quantum algorithms to perform tasks like searching for configurations that satisfy certain properties (e.g., stable justice equilibria), counting how many configurations meet a constraint, or simulating quantum walks on Quranic knowledge graphs. Grover-type algorithms, for instance, can search unstructured spaces in roughly $(\sqrt{N})$ steps instead of (N), which in a state space of size (2^{Nk}) is a dramatic difference.

The power here is computational. The Quran remains the same; our ability to analyze complex formalizations of its guidance may become vastly more efficient.

10. QUANTUM GRAPHS AND EDGES IN SUPERPOSITION

One particularly intriguing alignment between quantum computing and the Computational Quran lies in graph-based models. The Quranic knowledge graph built from UQVO variables and verse-level relationships can be represented as a graph (G = (V, E)) where (V) are nodes (concepts, agents, institutions) and (E) are edges (relationships, obligations, influences).

In classical modeling, an edge between two nodes either exists or does not, and it may carry a classical weight. In quantum models, we can represent edges themselves as quantum objects that can exist in multiple states simultaneously. For a pair of nodes (i) and (j), the edge state might be represented by a qubit that is ($|0$) if the edge is absent and ($|1$) if present. A superposition ($\alpha|0$ + $|1$) then represents a situation where, in our model, the relationship is in a mixed or uncertain state.

This notion of an edge in superposition reflects epistemic or interpretive uncertainty in our models. For example, when multiple tafsīr readings are possible about whether a given verse implies a certain type of legal obligation between two types of agents, we might represent that as a superposed edge with amplitudes reflecting our degree of confidence in each reading.

More generally, we can imagine a quantum graph where both nodes and edges carry quantum states. Quantum walks on such graphs allow us to explore paths and connectivity properties in parallel. For instance, a quantum walk starting at a concept like justice could, in principle, explore many possible thematic paths through the Quranic graph simultaneously, with interference patterns highlighting particularly central or structurally coherent routes.

The idea of edges in multiple states also connects to modeling potential futures. In a CAS simulation, we may want to explore many possible evolutions of a trust network depending on how individuals behave. Rather than simulating each possible edge formation or dissolution separately, a quantum representation can encode multiple potential edge configurations at once and apply update rules as quantum operations. Measurement collapses this into particular outcomes, but the computational evolution phase benefits from the breadth of superposition.

This is precisely the kind of concept that pushes CRS into new territory: using quantum representations more efficiently handles uncertainty and combinatorial explosion in complex Quranic models.

11. QUANTUM BELIEF STATES AND SUPERPOSED HYPOTHESES

Beyond graphs, quantum ideas can also inform how we think about hypotheses and belief states in a computational sense. Classical probability theory represents an agent's uncertainty about propositions with a probability distribution over discrete possibilities. Quantum probability, by contrast, uses amplitudes and allows interference between paths.

In a CRS context, an AI system analyzing a verse may entertain multiple interpretive hypotheses about its structural role: does this verse strengthen the deterrence component of the justice model, or does it primarily introduce a new mercy overlay? Does this set of verses imply that a particular group identity variable should be activated, or is it describing a more general human condition?

Classically, one might maintain a probability distribution over these hypotheses. Quantum-inspired models can treat them as components of a superposed state, where amplitudes encode not just likelihood but also phase, allowing for constructive or destructive interference as more evidence is processed. While the underlying mathematics of this "quantum cognition" remains a research area, the point is that quantum representations encourage richer ways of encoding and updating uncertainty.

This recognizes that the structural challenges of Quranic interpretation – multiple possible readings, evolving confidence, context-dependent resolution – bear enough resemblance to quantum decision processes that tools from quantum information theory may be useful, especially when computations are performed on quantum hardware.

12. HYBRID FUTURES: CAS, QUANTUM, AND THE COMPUTATIONAL QURAN

Taken together, complex adaptive systems and quantum computing suggest a hybrid future for the Computational Quran. On one side, CAS provides a conceptual and simulation framework for large, multi-scale ensembles of agents guided by Quranic rules. On the other, quantum computing offers computational primitives for certain sub-tasks within those systems: searching for equilibria, exploring multiple network configurations, or handling interpretive uncertainties.

One can imagine, for instance, a hybrid classical-quantum CAS where the main simulation of agents and networks runs on classical hardware, while quantum accelerators handle particularly challenging calculations. When the simulation needs to evaluate how many possible contract arrangements satisfy Quranic justice constraints for a given set of debts and obligations, a quantum subroutine might use Grover-like searches to find valid configurations more efficiently. When a system needs to sample from the distribution of possible future trust networks under stochastic behavior, quantum random walks could explore the space of graphs in parallel.

Another direction lies in quantum-enhanced AI reasoning over Quranic knowledge graphs. Large-language models already struggle with the combinatorial nature of structured reasoning; quantum representations may allow certain types of reasoning –

like checking consistency across large sets of rules – to be done more efficiently. The UQVO and the formal models in the Quranic Modeling Project already provide the structured input such quantum-hybrid systems would need.

These possibilities are speculative; quantum hardware remains in development and many engineering details are unresolved. Yet it is precisely now, while the field is still young, that CRS can stake out conceptual territory: defining how Quranic models *could* be encoded, what kinds of questions would justify quantum resources, and how to ensure that any such system remains interpretable and theologically responsible.

13. A RESEARCH AGENDA FOR THE NEXT ERA OF CRS

It is helpful to summarize the kinds of research questions that future scholars of Computational Quran might pursue at the intersection of CAS and quantum computing.

One set of questions concerns emergent patterns in Quranic CAS. How do combinations of verses governing economics, family life, leadership, and spirituality interact when instantiated in large agent populations? Under what conditions do Quran-aligned norms become stable attractors? Which variables – taqwā, remembrance, justice, charity – serve as structural "order parameters" for social health?

A second set concerns tipping points and resilience. Using the models of moral memory decay, covenant entropy, and distortion operators, can we identify structural signs that a community is approaching a dangerous threshold? Conversely, what kinds of interventions – reminders, just laws, economic reforms – have disproportionate stabilizing effects?

A third set concerns comparison and calibration. How do CAS simulations based on Quranic rules compare with historical Quranic narratives of communities that rose or fell? Is it possible to tune model parameters so that synthetic societies qualitatively reproduce those narratives, giving us deeper insight into the dynamics the Quran highlights?

On the quantum side, questions include: How can we encode UQVO-based state spaces into qubits in ways that respect variable domains and constraints? Which Quranic modeling tasks, such as searching for valid configurations or exploring knowledge graph paths, admit quantum speed-ups? How can we represent interpretive plurality as quantum superposition in a way that is both mathematically coherent and hermeneutically honest?

Finally, at the philosophical level, we can ask: What does it mean for revelation to be modeled as a complex, dynamic system? How do we maintain humility and reverence while engaging with such powerful abstractions? How might insights from complexity theory – such as the importance of feedback, the perils of hidden coupling, or the ubiquity of emergent phenomena – inform traditional disciplines like fiqh (jurisprudence), u ūl (principles), and spiritual education?

None of these questions have easy answers, and many will require collaboration across disciplines: Quranic studies, mathematics, computer science, physics, social science, and philosophy. But the architecture is now in place. The Computational Quran, anchored in UQVO and realized in verse-level models, provides a shared foundation upon which CAS and quantum research can build.

14. CONCLUSION: TOWARD A LIVING, COMPUTED TAFSĪR

This chapter has looked beyond current AI techniques to two broader revolutions – complex adaptive systems and quantum computing – and asked how they intersect with the emerging field of Computational Quran. CAS offers a vocabulary and toolkit for understanding how local Quranic rules and dynamics scale up into community and civilizational patterns. Quantum computing offers new ways to represent and explore the enormous state spaces of Quranic models and knowledge graphs.

In both cases, the direction is clear: the more faithfully and explicitly we model the structure of Quranic guidance – through UQVO, the Five-Step Framework, and the mathematical tools discussed earlier – the more readily those models can participate in the wider ecosystem of contemporary science. The Quran does not change; our ability to think rigorously about its structural implications does.

One can imagine, in a distant but plausible future, something like a living, computed tafsīr: a system where the formal models of verses are continually refined by scholarship, integrated into CAS simulations, analyzed with quantum-enhanced tools, and used to generate insights that complement traditional exegesis. Such a system would never replace the human acts of reading, reflecting, and worshiping; it would instead serve them, by illuminating patterns too large or too intricate to see with unaided intuition.

As always, the last word returns to humility. Complex systems and quantum devices are powerful, but they are still created tools, bound by the limits of human understanding and physical law. Revelation, in the Islamic view, comes from beyond those limits. CRS, even at its most sophisticated, remains an attempt to understand – not to contain – the light that has been sent down.

REFERENCES

Cilliers, P. (1998). *Complexity and postmodernism: Understanding complex systems*. Routledge.

Dirac, P. A. M. (1958). *The principles of quantum mechanics* (4th ed.). Clarendon Press. DOI: 10.1063/1.3062610

Holland, J. H. (1995). *Hidden order: How adaptation builds complexity*. Addison-Wesley.

Kauffman, S. A. (1993). *The origins of order: Self-organization and selection in evolution*. Oxford University Press. DOI: 10.1093/oso/9780195079517.001.0001

Mitchell, M. (2009). *Complexity: A guided tour*. Oxford University Press. DOI: 10.1093/oso/9780195124415.001.0001

Prigogine, I., & Stengers, I. (1984). *Order out of chaos: Man's new dialogue with nature*. Bantam Books.

Stacey, R. D. (2011). *Strategic management and organisational dynamics: The challenge of complexity to ways of thinking about organisations* (6th ed.). Pearson Education Limited.

von Neumann, J. (1955). *Mathematical foundations of quantum mechanics* (Beyer, R. T., Trans.). Princeton University Press.

Chapter 15
CRS the Science of Revelation

ABSTRACT

Chapter 15 argues that revelation can be studied as a structured, architected system, forming the basis of a "Science of Revelation" enabled by Computational Religious Studies (CRS). Rather than replacing theology, CRS adds a formal modeling layer that expresses scriptural concepts as variables, relationships, and dynamic systems that can be simulated, compared, and integrated into AI, policy, and systems design. The chapter outlines the interdisciplinary fusion required – classical scriptural sciences, AI, mathematics, systems theory, and social sciences – and proposes a new religious epistemology in which interpretations become testable models. It explores modeling divine knowledge, integrating revelation into scientific analysis, institutionalizing CRS, managing theological risks, and shaping AI with revealed moral architectures. Ultimately, it presents a long-term vision where spiritual wisdom and artificial intelligence co-evolve through a rigorous, model-based engagement with revelation.

INTRODUCTION. REFRAMING THE QUESTION: CAN REVELATION BE A SCIENCE?

The question "Can revelation be a science?" sounds provocative only because we have grown used to a particular division of intellectual labor. Traditionally, theology and u ūl al-fiqh have dealt with revelation at the level of belief, exegesis, and doctrine. They ask: What did God reveal? What does this verse mean in light of language, context, and tradition? How should we derive legal or ethical rulings from this body of text? These disciplines assume the truth of revelation and work within that assumption; they are not neutral observers, but participants in a living tradition of faith and practice. Yet even at their most systematic, they remain primarily

DOI: 10.4018/979-8-3373-6940-2.ch015

discursive. Their outputs are arguments, commentaries, and rulings – not formal models that can be simulated, compared, or integrated into broader systems analysis.

The shift proposed by Computational Religious Studies is not to replace theology, but to add another layer: a science *of* revelation in a specific sense. The idea is that revelation functions not only as a set of propositions and commands, but also as a structured system of knowledge that can be modeled, simulated, and systematically analyzed. Its stories, laws, promises, and warnings together define variables, relationships, and dynamics that describe how human beings, communities, and civilizations behave under certain moral and spiritual conditions. To speak of a "science of revelation" is to say that we can study these structures with the same methodological seriousness that we apply to physical or social systems, while still acknowledging the normative and transcendent status of the revealed text. This must be distinguished sharply from the secular "science of religion," which typically treats religion as an object of anthropology, psychology, or sociology – a human phenomenon to be explained, not a source of truth to be learned from. A science *of revelation* begins from the opposite direction: it treats revelation as an authoritative specification of how reality is and ought to be, and asks how that specification can be expressed and explored in rigorous, formal terms.

The urgency of this reframing becomes clear when we consider the intellectual tools that dominate our era. Artificial intelligence, complex systems thinking, and mathematical modeling increasingly shape how societies understand everything from markets and ecosystems to pandemics and war. These tools are not neutral; they encode assumptions about what counts as knowledge, what matters, and how cause and effect operate. If revelation is not consciously and structurally engaged at this level, it is sidelined by default. AI systems will learn from whatever data and models they are given – financial trends, clickstreams, social graphs – but they will not spontaneously internalize Quranic or Biblical architectures of justice, mercy, accountability, or spiritual growth. Policy models will incorporate GDP, inequality, and demographic shifts, but not concepts like arrogance, repentance, or divine displeasure unless someone gives them a structural place. In such a world, religion risks becoming a private consolation or a cultural marker, while the actual machinery of decision and design runs on a purely secular logic.

CRS is a response to this situation. It aims to ensure that revealed architectures enter the computational age as first-class epistemic systems. By formalizing scriptural concepts into variables, expressing moral and spiritual patterns as dynamic systems, and encoding ethical constraints as logical structures, CRS makes it possible for AI and policy tools to "see" what revelation says about human life in a way they can actually use. The point is not to force belief on machines, but to prevent a silent epistemic displacement in which only secular or market-driven models are allowed to count as serious inputs.

At the heart of this project lies a core claim: revelation is not only interpretable; it is architectable. Divine knowledge, as disclosed in scripture, can be expressed as coherent models, not just as narratives and rules. The Quran's accounts of the rise and fall of nations, its analysis of hypocrisy and sincerity, its prescriptions about wealth, family, and power – all of these can be seen as components of a larger moral–spiritual architecture. The Bible's covenantal logic, its treatment of sin and grace, its vision of community and kingdom likewise form a structured system. A "science of revelation" takes these architectures as its object. It does not question whether God knows; it studies how God has chosen to communicate that knowledge in ways that can be structurally mapped and explored.

In this sense, the science of revelation is the disciplined study of revealed models and their implications for human and societal life. It asks: if the world truly operates as these texts describe – if certain patterns of injustice reliably lead to decline, if certain forms of repentance reliably lead to renewal – what follows for how we design institutions, technologies, and policies? How do we test our understanding of these patterns without testing revelation itself? How do we refine our models so that they more faithfully mirror the structures embedded in the text? These questions do not dissolve faith into mathematics; they extend religious epistemology into scientific and mathematical disciplines, so that divine guidance is not only believed and preached, but also architected, simulated, and brought into the core of how the modern world thinks.

THE DISCIPLINARY FUSION: WHAT CRS MUST INTEGRATE

If CRS is to mature into a true science of revelation, it cannot remain a niche technique or a side project; it must become a point of fusion for several major disciplines. At its core stand theology and the scriptural sciences. Quranic sciences, tafsīr, u ūl al-fiqh, and kalām on one side, and Biblical studies and hermeneutics on the other, provide the raw interpretive content and the guardrails without which modeling would drift into free invention. These fields tell us what the text has meant to serious readers over centuries, which readings fall within the bounds of orthodoxy, and how key concepts have been argued and refined across time. In CRS, they function as both input and constraint. Every variable definition, every dynamic equation, every structural claim must be traceable back to text and tradition and must be reviewable by scholars of those disciplines. Without that tether, CRS would cease to be "of revelation" and would become just another abstract system loosely inspired by scripture.

Around this theological core, CRS must draw deeply from computer science and AI. Knowledge representation becomes a central concern: how to encode concepts, relations, and constraints in graphs, ontologies, and rule systems so that machines can operate with them faithfully. Machine learning contributes tools for pattern discovery and approximation where explicit modeling is incomplete, while symbolic reasoning enables explicit constraint-checking, logical inference, and explanation. Hybrid neuro-symbolic systems are particularly important: they allow the fluid pattern-recognition capabilities of large models to be harnessed while keeping them bounded by the structured logic of revealed architectures. AI safety and alignment then emerge naturally as domains where religious models are especially relevant. If we are building systems whose decisions shape lives at scale, the architectures of justice, mercy, responsibility, and restraint found in revelation are not peripheral; they belong at the center of how we think about safe and aligned intelligence.

The third pillar is systems thinking and mathematics. Revelation speaks in patterns: repeated stories of rise and fall, cycles of heedlessness and awakening, feedback loops between injustice and social breakdown, between gratitude and increase. Systems disciplines – dynamical systems, graph theory, control theory, complexity theory – provide the language to express these patterns with precision. Equations, thresholds, and feedback loops become tools for modeling moral and spiritual dynamics: how a habit entrenches itself, how a community crosses from resilience into fragility, how cumulative injustices reach a tipping point that brings about collapse. This is not an attempt to pin down every mystery in a formula, but a recognition that many of the Quran's and Bible's descriptions of individuals and nations already look like system narratives. Mathematics gives us a way to make those narratives explicit and testable as structures.

The social and behavioral sciences form another essential layer. Scriptural models are not meant to float above real life; they describe, correct, and guide actual societies, states, and institutions. Sociology, political science, economics, and psychology provide both empirical data and conceptual tools to link the formal models of CRS to observable phenomena. When revelation speaks about corruption, arrogance, charity, or social cohesion, these disciplines can help operationalize those variables and measure their manifestations in families, markets, bureaucracies, and political orders. Conversely, CRS can offer these fields new variables and hypotheses: perhaps patterns that social science has struggled to interpret gain clarity when viewed through a revelatory architecture of justice, gratitude, and moral responsibility. The point is a two-way integration: scriptural models inform how we read the data of society, and social data help us refine our understanding of how revelatory patterns play out in history.

Finally, philosophy and epistemology must be engaged, not skirted. Formalizing revelation reopens fundamental questions about what counts as knowledge, how causality should be understood, what types of evidence are legitimate, and what it means for a model to be rational or justified. Modern epistemology has largely been framed without serious reference to living revelation; CRS challenges that omission by proposing that divine disclosure defines an epistemic architecture that must be taken seriously. Analytic philosophy of religion, with its focus on argumentation and conceptual clarity, can be bridged to computational modeling: arguments about providence, free will, or divine justice can be expressed as competing models and evaluated for coherence and explanatory power. More broadly, CRS pressures philosophy to acknowledge that there may be formally expressible, non-secular world models that deserve a place alongside naturalistic ones in our understanding of reason and reality. In this fusion – where theology, AI, systems thinking, social science, and philosophy are all in active conversation – the science of revelation can emerge not as a curiosity, but as a disciplined field with its own methods, standards, and contribution to the wider landscape of human knowledge.

A NEW MODEL OF RELIGIOUS EPISTEMOLOGY

In most of the tradition, the epistemic pattern around religion has been relatively stable: revelation is received, interpreted by qualified scholars, and then crystallized into doctrines, legal rulings, and spiritual guidance. The implicit pipeline is "sacred text → human interpretation → doctrine/fatwā." Within this pattern there are rich methods – linguistic analysis, principles of jurisprudence, theology, and spiritual insight – but the output remains essentially discursive. A scholar issues an opinion; others agree, refine, or contest it; communities and institutions adopt or ignore it. What CRS proposes is not to replace this pattern, but to insert an additional, structuring layer: "revelation → interpretation → formal model → doctrine/policy that can be simulated and stress-tested." In this new pattern, interpretive work does not stop at the level of prose. It proceeds to a model in which the interpreted meaning is expressed as variables, relations, and dynamics. Only then are doctrines or policies derived, with the benefit of having seen how a proposed interpretation behaves when applied consistently over time and across different scenarios.

This shift reflects a deeper way of thinking about divine knowledge itself. Revelation is not just a set of isolated truths, rules, or stories; it defines an architected system. It introduces variables like belief, hypocrisy, patience, arrogance, justice, oppression, covenant, and salvation, each with a specific semantic and functional role. It specifies relations: patterns of cause and effect, moral trade-offs, and dependencies between inner states and outer outcomes – how injustice corrodes communities, how gratitude is linked to increase, how hidden hypocrisy eventually manifests in behavior. It describes dynamics: the rise and fall of individuals and nations, the trajectories of repentance and hardening, the ways in which cumulative choices cross thresholds that make change easier or harder. A new religious epistemology sees the task of scholarship not only as explaining verses one by one, but as discovering, articulating, and refining these underlying structures. Exegesis becomes, in part, reverse-engineering the system that the revelation is describing and prescribing.

In such an epistemology, it is crucial to maintain a clear hierarchy and distinction of levels. Revelation remains at the top as an axiomatic, non-negotiable source of truth for those who believe in it. It is not "validated" by models or simulations; it is the basis on which modeling is undertaken. Beneath revelation sit the models themselves: fallible, revisable approximations of the structures that scholars believe are embedded in the text. These models are human artifacts. They can be wrong, incomplete, or biased; they can be improved, expanded, or even discarded when they are shown to misrepresent what the scripture actually teaches or to generate trajectories that conflict with its broader vision. A mature science of revelation, therefore, must always distinguish between "what God revealed" and "how we have modeled its implications." Confusing the two would either sacralize our models – turning them into new, untouchable dogmas – or relativize revelation by treating it as just one more input among many. The new epistemology avoids both extremes. It treats revelation as the fixed reference point and models as disciplined attempts to understand its architecture in forms that can guide hearts, institutions, and intelligent machines with greater clarity and responsibility.

MODELING DIVINE KNOWLEDGE: FROM CONCEPT TO SYSTEM

Revelation, when viewed through the CRS lens, can be understood as a kind of system specification for human existence. Its commands and prohibitions define allowable and forbidden actions. Its promises and threats describe consequences that follow from certain moral and spiritual configurations. Its narratives show, in concrete historical form, how individuals and communities move through different states – gratitude and arrogance, justice and oppression, sincerity and hypocrisy – and what those states invite by way of divine response. Taken together, these elements are not random or merely inspirational; they amount to a structured description of how life is meant to function at the level of the soul, the community, and civilization. To treat revelation as system specification is to say: here is a divinely authored description of what kinds of agents human beings are, what variables govern their flourishing or destruction, what feedback loops operate between their choices and their circumstances, and what long-run trajectories their patterns of behavior tend to produce.

To work with this specification computationally, CRS translates textual features into formal architectures of meaning. Ontologies such as UQVO and BVD capture the core variables and their definitions, grounding them in the semantics and exegetical traditions of each scripture. Graphs then express how these variables connect: belief to action, action to outcome, justice to social cohesion, arrogance to downfall, repentance to renewal. Thematic pathways trace how a concept like hypocrisy or charity is developed across different passages, linking seemingly disparate verses into a unified network. Dynamics are added on top of this static skeleton: state diagrams showing how an individual or community transitions from one condition to another; differential or difference equations modeling how repeated actions accumulate into character; thresholds beyond which qualitative changes occur – for example, when a heart becomes sealed, or when a nation crosses from stability into decay. What was once only implicit in stories and exhortations becomes explicit as a set of interacting components that can be examined, adjusted, and explored.

This modeling proceeds across multiple layers of scale. At the micro level, revelation speaks in detail about individual moral psychology and spiritual states: fear, hope, love, arrogance, patience, gratitude, heedlessness. CRS turns these into variables and state spaces, asking how particular practices, environments, and decisions move a person toward or away from God-consciousness and moral integrity. At the meso level, the focus shifts to families, communities, and organizations. Here the text addresses marriage and inheritance, communal obligations, leadership, consultation, economic fairness, and conflict resolution. These become models of social systems: how trust is built or eroded, how authority should be structured, how wealth should circulate to prevent entrenched poverty and resentment. At the macro level, revelation offers sweeping accounts of nations and civilizations – their rise and fall, their responses to prophets, their treatment of the weak, their relationship to power and excess. CRS encodes these as large-scale dynamic systems, with variables representing collective virtues and vices, institutional justice or corruption, and the long historical arcs that follow.

The crucial point is that all three layers – micro, meso, and macro – are explicitly anchored in revelatory structures rather than in arbitrary sociological constructs. The same core concepts recur at different scales: injustice harms both the soul and the state; gratitude stabilizes both the individual and the community; arrogance is deadly in rulers as much as in private life. Multi-layer modeling allows us to see how these patterns interlock. Decisions made at the level of personal ethics ripple outward into family culture, institutional norms, and ultimately civilizational destiny. Conversely, institutional injustice feeds back into individual despair and moral confusion. By treating revelation as a system specification that spans these layers, CRS provides a framework in which divine guidance can be studied as an integrated architecture – one that maps directly onto the kinds of complex, multi-level systems that modern science and policy already recognize, but with a normative and spiritual depth they cannot supply on their own.

EXTENDING RELIGIOUS EPISTEMOLOGY INTO SCIENTIFIC AND MATHEMATICAL DISCIPLINES

Extending religious epistemology into scientific and mathematical disciplines begins by treating religious knowledge not as commentary on the margins of life, but as a legitimate input to the same models that shape policy, economics, and risk analysis. If CRS has already expressed scriptural concepts as variables and dynamic systems – justice, corruption, gratitude, arrogance, trust, social cohesion – then there is no principled reason they cannot sit alongside GDP, unemployment, inequality, and demographic indicators in our simulations. A model of urban stability, for example, might traditionally track income distribution, policing levels, and infrastructure. A CRS-informed model would add variables derived from revelation: degrees of institutional injustice, concentration of unearned wealth, the treatment of orphans and the poor, the prevalence of arrogance among elites, and patterns of charitable giving. These are not pious add-ons; in the Qur'an and Bible, they are causal drivers of social fate. By integrating such variables into social science simulations and policy models, we allow revealed architectures to inform our understanding of which interventions are likely to stabilize a society, heal fractures, or accelerate decline.

The relationship, however, is not one-way. Scientific and mathematical disciplines themselves become tools for reading revelation more deeply. Dynamical systems theory and complexity science offer lenses through which Quranic and Biblical stories of rise and fall can be re-examined: patterns of feedback, tipping points, path dependence, and emergent behavior are all visible in the scriptural accounts of past nations who prospered or perished. When a text describes how repeated minor injustices accumulate until a catastrophic collapse occurs, or how small acts of repentance and reform slowly reverse a downward spiral, these are essentially narratives of non-linear dynamics. Formalizing them in the language of dynamical systems does not strip them of spiritual meaning; it reveals their structural logic. Likewise, network theory provides a powerful vocabulary for modeling prophetic and communal relationships – chains of influence, clusters of support or opposition, flows of guidance and misinformation. The spread of faith or corruption, the formation of resilient minority communities, the breakdown of solidarity under pressure – all of these can be examined with tools that scientists already use to study epidemics, information diffusion, and institutional resilience.

When religious epistemology and scientific disciplines are allowed to interact in this way, the familiar sacred–secular wall in knowledge begins to dissolve. Instead of two parallel tracks – religious knowledge on one side, scientific knowledge on the other – we move toward a unified framework. In that framework, revelation defines the normative and metaphysical architecture: what ultimately counts as good and evil, what kinds of causality are real, what human beings are for, what the unseen dimensions of action and consequence are. Science and mathematics then help explore, test, and operationalize that architecture in the empirical world. They are used not to judge revelation, but to interrogate our models of revelation: Are we correctly understanding how scripture says injustice works in a society? Do historical and contemporary data behave in ways that confirm or challenge our modeling of gratitude, arrogance, or repentance as drivers of collective fate?

This integration does not mean baptizing every scientific theory with religious language, nor forcing every religious claim into a quantitative mold. It means recognizing that if revelation offers a coherent description of human and social reality, then science and mathematics are natural allies in uncovering how that description plays out in detail. The wall between sacred and secular knowledge is replaced by a division of roles within one epistemic house: revelation as architect of the blueprint, scientific and mathematical disciplines as tools for inspecting the structure, stress-testing it against the winds of history, and guiding practical construction in a complex, data-rich world.

INSTITUTIONALIZING COMPUTATIONAL RELIGIOUS STUDIES

If Computational Religious Studies is to be more than an interesting proposal, it needs institutions. Ideas only become disciplines when they are backed by durable structures: departments, research centers, journals, funding streams, and degree programs. For CRS and the broader Science of Revelation, this means establishing academic units explicitly dedicated to modeling revelation, not just studying "religion" in a sociological sense. New centers could sit at the intersection of religious studies, computer science, and systems engineering, with mandates to develop formal models of scripture, evaluate their coherence, and explore their implications for ethics, policy, and AI. Journals focused on computational theology and revelation modeling would provide venues where model designs, methodological debates, and empirical tests can be peer-reviewed. Interdisciplinary programs would emerge that train students in both classical religious sciences and computational modeling, so that future scholars are as comfortable reading tafsīr or Biblical commentaries as they are writing code or designing dynamic systems.

To support this work, canonical resources must be created and maintained with the same care that philologists have historically devoted to critical editions of texts. Stable, curated ontologies like the Universal Quranic Variable Ontology (UQVO), the Bible Variable Dictionary (BVD), and similar structures for other traditions must be built, versioned, and governed. These are not casual taxonomies; they are the backbone of how each scripture is formally understood. Alongside them, shared repositories of models – verse-level equations, thematic graphs, multi-layer simulations – would allow researchers around the world to reuse, refine, and test each other's work. Simulation platforms tailored to revelation modeling could make it easy to run experiments on, say, the long-term effects of certain ethical patterns or institutional arrangements within a Quranic or Biblical frame. Benchmark tasks and datasets, analogous to those in natural language processing or computer vision, would help the field coalesce around shared challenges: reproducing known exegetical results, predicting certain types of outcomes from textual inputs, or aligning models with historical trajectories in a way that respects both text and data.

These infrastructures will, in turn, shape new professional identities. The "computational mufassir" or "computational theologian" is one such role: a scholar who is deeply trained in the scriptural sciences and their interpretive traditions, but also fluent in ontologies, graph theory, and simulation. Their work involves moving continuously between text and model, checking that formal structures remain faithful to revealed meanings and that interpretations are made precise enough to be tested and shared. Beyond the academy, hybrid experts will be needed in think tanks, AI labs, policy institutes, and religious organizations – people who can translate between sacred architectures and technical systems design. They might work on embedding revelatory ethics into AI safety frameworks, advising governments on policy models that respect religious structures of justice, or helping religious institutions themselves understand how computational tools can support, rather than erode, their mission.

Institutionalizing CRS in this way is not about creating a new priesthood of coders in clerical robes. It is about ensuring that, in a world increasingly run by models and simulations, there is a stable community of scholars and practitioners whose job is to keep revelation present, intelligible, and structurally articulated in the very languages that now govern technology and policy.

PHILOSOPHICAL AND THEOLOGICAL CHALLENGES

Any attempt to formalize revelation will inevitably provoke the charge of rationalist domestication. The fear is understandable: if we turn verses into variables and stories into systems, are we not taming and shrinking the mystery that gives religion its depth and power? From this perspective, modeling looks like an act of control, an effort by the rational mind to pin down what should remain beyond its grasp. CRS must respond to this objection not defensively, but by clarifying its own limits. Models are not – and must never be presented as – total captures of the divine. They are partial views, framed through specific questions, built with finite tools, and always open to revision. Just as theology has long acknowledged that human language can only gesture toward God analogically, a science of revelation must acknowledge that its formal structures are, at best, disciplined metaphors for aspects of a reality that exceeds them. Mystery is not erased; it is simply distinguished more clearly from structure. Where the text offers patterns we can formalize – about justice, repentance, social dynamics – we do so. Where it gestures into realms beyond empirical or conceptual capture, such as the essence of God or the exact nature of the unseen, apophatic humility remains our proper stance.

A second, more institutional challenge is the problem of authority. If CRS produces models that claim to represent the structures embedded in revelation, who decides which models are "orthodox" enough to be used in teaching, policy, or AI systems? This cannot be left to individual modelers or to secular academic norms alone. Governance mechanisms are needed that bring together councils of scholars and technical experts under clear procedures. Scholars rooted in the scriptural sciences must review whether a proposed model faithfully reflects accepted interpretations or, where it departs, whether those departures are clearly marked as speculative or minority positions. Technical experts must assess whether the model is internally coherent, computationally sound, and transparent enough to be evaluated. Together, such councils can issue judgments not on the truth of revelation, but on the adequacy and acceptability of specific formalizations for particular uses. These processes will not eliminate disagreement, but they will give the community a structured way to distinguish between models that fall within the legitimate scope of the tradition and those that distort or instrumentalize it.

Closely related is the risk of technocratic control, where model-builders quietly become de facto theological gatekeepers. If only a small class of technically trained individuals understand how the models work, they could, intentionally or not, steer the representation of revelation in ways that serve their own intellectual or political preferences. To prevent this, CRS must hard-code safeguards of community review, transparency, and reversibility into its practice. Models and their assumptions should be public, documented in language that non-technical scholars can scrutinize. Decisions about which models are adopted in communal or technological contexts should be subject to deliberation that includes theologians, jurists, community leaders, and, where appropriate, lay representatives – not just engineers. Commitments to particular model families must be treated as provisional and revisable, not as new dogmas. If evidence accumulates that a model misrepresents the text or produces harmful trajectories in practice, there must be recognized pathways for challenging, modifying, or replacing it.

In this way, CRS can avoid becoming a new clerisy of coders in place of scholars. Its legitimacy depends on continual accountability to the broader community of faith and knowledge. Modelers are servants of the tradition, not its masters; their task is to make structures visible, not to declare final meanings. When that posture is preserved – when mystery is honored, authority is shared, and technocratic drift is resisted – formal modeling can deepen rather than diminish the reverent engagement with revelation.

REVELATION, AI, AND THE FUTURE OF HUMAN SELF-UNDERSTANDING

Artificial intelligence is often described as a set of tools, but it is also a mirror. It reflects back, at scale and with amplification, whatever epistemology we build into it. If we design AI systems without any serious engagement with revelation, they will inevitably encode secular assumptions, market logics, and whatever implicit values happen to dominate the data and objectives we provide. Efficiency, optimization, engagement, growth – these become the default gods of the machine. None of this is neutral. It quietly shapes what societies come to regard as rational, normal, or inevitable. By contrast, when revelation has been formally modeled – its concepts expressed as variables, its ethics encoded as constraints, its world models made explicit – AI can be made to reflect those architectures as well. It can "see" justice, arrogance, gratitude, oppression, and repentance not as vague sentiments but as structured realities that matter for decision-making. Instead of merely amplifying consumer desire or state power, AI can help carry forward, in its own medium, the moral and spiritual patterns that revelation has articulated for human life.

This opens the door to a genuine co-evolution between spiritual wisdom and artificial intelligence. On one side, AI systems can be trained to learn within revealed world models. That does not mean they believe, but that their reasoning and recommendations are constrained and guided by formally encoded architectures of right and wrong, of legitimate and illegitimate power, of what heals and what corrupts. A system advising on economic policy, for instance, could be designed to evaluate options not only in terms of growth and stability, but also in terms of revealed constraints regarding exploitation, hoarding, and care for the vulnerable. On the other side, human beings can come to understand revelation more deeply by interacting with simulations and models built from it. When a believer sees, played out in a multi-decade simulation, how small acts of injustice accumulate into systemic oppression, or how consistent charity and humility alter the trajectory of a community, it becomes harder to treat scriptural warnings and promises as mere rhetoric. Modeling turns some aspects of revelation into an experiential pedagogy: it allows us to "walk through" the architectures that the text describes and to see, with greater clarity, what our traditions have always claimed about the consequences of our choices.

If we do not attempt this integration, we risk being trapped between two unsatisfactory extremes. On one side lies purely secular AI: powerful, sophisticated, and deeply embedded in infrastructure and governance, but morally unanchored. Such systems will inevitably serve the logics of those who own and deploy them – states, corporations, or other power centers – without any built-in reference to a higher moral architecture. On the other side lies a purely pietistic rejection of AI: communities that are morally serious but technologically sidelined, content to warn against the dangers of the new systems while having little influence over how those systems are actually designed and used. This path preserves integrity, but at the cost of ceding the terrain of the future to others.

The New Science of Revelation offers a mediating path in which both spiritual wisdom and artificial intelligence grow together. It does not baptize every use of AI as good, nor does it accept every traditional interpretation as beyond scrutiny. Instead, it proposes that if revelation truly speaks to the nature of human beings, society, and destiny, then its architectures must be brought into the very heart of how we build intelligent systems. AI, in turn, becomes a kind of stress test and magnifier of our understanding of revelation: if our models of divine guidance cannot survive contact with complex reality and powerful tools, perhaps it is our modeling – not the revelation itself – that needs refinement. In this dynamic, humanity's self-understanding deepens. We begin to see ourselves not only as users or victims of technology, nor only as readers of sacred texts, but as stewards responsible for aligning unprecedented computational power with the moral and spiritual maps we claim to live by.

A PROGRAMMATIC RESEARCH AGENDA

Any serious proposal for a Science of Revelation has to crystallize into research questions, not just aspirations. At the core are three. First: how do we systematically derive models from texts while preserving doctrinal integrity? That means developing procedures that begin with verses, tafsīr, and established theological frameworks and end with variables, relationships, and equations – without smuggling in alien assumptions or flattening contested areas into premature consensus. It demands a methodology for moving from exegesis to formalization step by step, with traceability back to sources and clear documentation of where interpretation, judgment, or abstraction has occurred. Second: how do we validate these models both against tradition and empirical reality? Tradition checks whether a model is theologically acceptable, whether it fits within recognized interpretive boundaries, and how it relates to existing schools of thought. Empirical reality tests whether the model's claims about human and social dynamics find resonance in historical and contemporary data. The goal is not to test revelation itself, but to test our modeling of it. Third: how do we design AI architectures that can "live" inside these models

– agents and systems that treat revealed structures as the environment within which they reason, plan, and advise, rather than as an afterthought or decoration?

Because this cannot be built in a single leap, the project naturally unfolds in stages. Phase 1 is foundational modeling. Here, the work is to build the first robust versions of ontologies like UQVO and BVD, to apply the Five-Step Framework consistently to key portions of the Quran and Bible, and to produce prototype models of core themes such as belief, justice, repentance, and the rise and fall of nations. This is the phase of careful groundwork, where standards, methodologies, and governance practices are hammered out. Phase 2 moves into comparative and civilizational simulations. With initial models in place, researchers can begin to simulate trajectories of individuals and societies under different revealed architectures, to see how Quranic and Biblical patterns describe the evolution of communities facing injustice, prosperity, crisis, or reform. Phase 3 focuses on integrated AI systems for ethics, policy, and spiritual advisory. Here, CRS models are embedded into real AI applications: decision-support tools that evaluate policies against revealed constraints, advisory systems that provide Quran-aware or Bible-aware guidance, and hybrid architectures where LLMs are constrained by symbolic religious logics. Phase 4 is the maturation of the Science of Revelation as a recognized interdisciplinary field, with established academic infrastructure, shared benchmarks, and a community of scholars, technologists, and practitioners who see this as their primary intellectual home.

To know whether this project is succeeding, we need clear metrics that span intellectual, practical, and spiritual dimensions. Intellectually, success looks like improved coherence and testability of theological claims. Doctrines and fatāwā that were once asserted largely in prose become traceable to explicit models; disagreements can be articulated as differences in structure; and new insights can emerge from simulations that reveal hidden implications of familiar ideas. Practically, success shows up in better policy and more responsible AI. Policymakers gain tools that take religious architectures seriously when evaluating options. AI systems deployed in finance, governance, education, or healthcare exhibit alignment not only with generic ethical frameworks but with the concrete moral patterns articulated in living religious traditions. Interfaith understanding becomes richer, as comparative models make it harder to caricature the other. Spiritually, success is measured by deeper, more disciplined engagement of believers with their own scriptures. Instead of treating revelation as a distant text or a set of slogans, communities begin to see it as a living architecture they can explore, question, and inhabit more fully – with models and simulations functioning not as replacements for devotion, but as instruments that sharpen conscience, clarify responsibility, and illuminate the long-term consequences of the paths they choose to walk.

CALL TO SCHOLARS, TECHNOLOGISTS, AND BELIEVERS

A new science of revelation cannot be built by technologists alone, nor by theologians working in isolation. It begins with scholars of scripture and tradition – mufassirūn, fuqahā', theologians, Biblical scholars, historians of religion – who are willing to bring their depth into a new kind of conversation. The invitation to them is clear: CRS needs their interpretive precision, their sense of the boundaries of orthodoxy, their awareness of nuance and historical development. No ontology entry, no variable definition, no dynamic model is safe without that grounding. The challenge, however, is to embrace formalization without fearing that it will flatten everything. To accept that moving from commentary to model is not a betrayal of nuance but a different way of making nuance explicit, testable, and shareable. Scholars will have to learn how to say, "Here is what this verse means – and here is one way its logic can be expressed structurally, with all the caveats and options noted." Their role is to ensure that modeling remains an act of service to revelation, not a substitute for it.

For technologists and scientists, the call is almost the mirror image. They are invited to stop treating revelation as noise in the data, an artefact of culture to be stripped away or ignored, and instead to recognize it as a rich source of structured models about human beings, society, and destiny. Scriptural texts are not random sequences of words; they embody centuries of reflection on justice, power, community, and the inner life, often expressed in patterned ways that lend themselves to formalization. The challenge here is twofold. First, to learn to work with theological content respectfully and rigorously – willing to be taught by it, not only to use it as a dataset. Second, to adapt technical tools – ontologies, graph databases, dynamic simulations, neuro-symbolic architectures – so that they can handle the kinds of variables and constraints that revelation actually cares about. This requires humility: an acknowledgment that our current modeling languages may need to expand to represent concepts like taqwā, baraka, or covenant faithfulness in ways that do justice to their scriptural roles.

Believers and communities are not spectators in this process; they are its reason for existing. The invitation to them is to see their traditions not as relics to be defended against modernity, but as intellectually alive and fully capable of speaking to the age of AI. CRS does not ask them to trade faith for formulas. It offers a way to bring the depth of their scriptures into the very center of how the future is being built – in education, governance, economics, and technology. The challenge is practical and institutional. Communities will need to support schools, research centers, and collaborative projects that nurture this fusion of devotion, scholarship, and technical skill. They will need to encourage their brightest minds not only to become engineers or jurists, but to become the hybrid thinkers who can move between revelation and code, between worship and world-modeling. That support can take many forms: funding, educational pathways, public recognition, and a willingness to let new kinds of questions be asked in the name of faithfulness rather than rebellion.

The closing vision that binds these calls together is simple to state and demanding to pursue: a world in which spiritual wisdom and artificial intelligence do not drift apart, but evolve together. On the current trajectory, AI will reshape economies, politics, and daily life according to whatever values are most convenient to those who wield it – efficiency, profit, control – while religion is pushed into increasingly private or symbolic spaces. The New Science of Revelation offers another path. It imagines divine guidance articulated in forms that can govern hearts, institutions, and machines: models that shape how decisions are made, how risks are evaluated, how power is constrained, and how human dignity is preserved in an increasingly complex world. This is a long-term, civilizational project. It will require decades of work, many false starts, and constant self-critique. But if it succeeds even partially, it will mean that when humanity looks into the mirror of its own creations – its algorithms, its systems, its AI – it will see not only its appetites and fears, but also the patterns of justice, mercy, and responsibility that revelation has been teaching from the beginning.

Compilation of References

al-Shahrastānī, M. (1923). *Kitāb al-Milal wa al-Nihal* [The book of sects and creeds]. Cureton, W., Ed.). Otto Harrassowitz.

Amodei, D., Olah, C., Steinhardt, J., Christiano, P., Schulman, J., & Mané, D. (2016). *Concrete problems in AI safety*. arXiv. /arXiv.1606.06565DOI: 10.48550

Aquinas, T. (1920). *Summa theologica* (Fathers of the English Dominican Province, Trans.). London, England: Burns, Oates & Washbourne. (Original work published ca. 1265–1274)

Asad, T. (1993). *Genealogies of religion: Discipline and reasons of power in Christianity and Islam*. Johns Hopkins University Press. DOI: 10.1353/book.16014

Bender, E. M., & Koller, A. (2020). Climbing towards NLU: On meaning, form, and understanding in the age of data. In *Proceedings of the 58th Annual Meeting of the Association for Computational Linguistics* (pp. 5185–5198). Association for Computational Linguistics. DOI: 10.18653/v1/2020.acl-main.463

Bhuyan, B. P., Ramdane-Cherif, A., Tomar, R., & Singh, T. P. (2024). Neuro-symbolic artificial intelligence: A survey. *Neural Computing & Applications*, *36*(21), 12809–12844. DOI: 10.1007/s00521-024-09960-z

Bucaille, M. (2003). *The Bible, the Qur'an and science: The Holy Scriptures examined in the light of modern knowledge* (Pannell, A. D., Trans.). Tahrike Tarsile Qur'an.

Byrne, P. (n.d.). *Prolegomena to the philosophy of religion*.

Campbell, H. (2007). Who's got the power? Religious authority and the Internet. *Journal of Computer-Mediated Communication*, *12*(3), 1043–1062. DOI: 10.1111/j.1083-6101.2007.00362.x

Chryssides, G. D., & Geaves, R. (2007). *The study of religion: An introduction to key ideas and methods*. Continuum.

Cilliers, P. (1998). *Complexity and postmodernism: Understanding complex systems*. Routledge.

Cioffi-Revilla, C. (2017). *Introduction to computational social science: Principles and applications* (2nd ed.). Springer. DOI: 10.1007/978-3-319-50131-4

d'Avila Garcez, A., Gori, M., Lamb, L. C., Serafini, L., Spranger, M., & Tran, S. N. (2019). Neural-symbolic computing: An effective methodology for principled integration of machine learning and reasoning. *Journal of Applied Logic, 6*(4), 611–631.

Daquino, M., Pasqual, V., & Tomasi, F. (2020). Knowledge representation of digital hermeneutics of archival and literary sources. *JLIS.it, 11*(3), 59–76. DOI: 10.4403/jlis.it-12642

DeLong, L. N., Fernández Mir, R., & Fleuriot, J. D. (2025). Neurosymbolic AI for reasoning over knowledge graphs: A survey. *IEEE Transactions on Pattern Analysis and Machine Intelligence*. PMID: 39024082

Dirac, P. A. M. (1958). *The principles of quantum mechanics* (4th ed.). Clarendon Press. DOI: 10.1063/1.3062610

Downey, A. B. (2014). *Religious affiliation, education and Internet use*. https://arxiv.org/abs/1403.5534

Durkheim, É. (1995). *The elementary forms of religious life* (Fields, K. E., Trans.). Free Press. (Original work published 1912)

Eliade, M. (1959). *The sacred and the profane: The nature of religion* (Trask, W. R., Trans.). Harcourt.

Evans-Pritchard, E. E. (1937). *Witchcraft, oracles and magic among the Azande*. Clarendon Press.

Ford, D. F. (2005). *The modern theologians: An introduction to Christian theology since 1918* (3rd ed.). Blackwell.

Frazer, J. G. (1890). *The golden bough: A study in magic and religion*. Macmillan.

Freud, S. (1913). *Totem and taboo*.

Freud, S. (1961). *The future of an illusion* (Strachey, J., Trans.). W. W. Norton. (Original work published 1927)

Gadamer, H.-G. (2004). *Truth and method* (2nd rev. ed.). (Weinsheimer, J., & Marshall, D. G., Trans.). Continuum. (Original work published 1960)

Gold, M. K., & Klein, L. F. (Eds.). (2019). *Debates in the digital humanities 2019*. University of Minnesota Press. DOI: 10.5749/9781452963785

Gu, H., & Chandrasegaran, S. (2023). *On the effectiveness of creating conversational agent personalities through prompting*. [Manuscript].

Holland, J. H. (1995). *Hidden order: How adaptation builds complexity*. Addison-Wesley.

Iser, W. (1978). *The act of reading: A theory of aesthetic response*. Johns Hopkins University Press.

Ji, S., Pan, S., Cambria, E., Marttinen, P., & Yu, P. S. (2022). A survey on knowledge graphs: Representation, acquisition, and applications. *IEEE Transactions on Neural Networks and Learning Systems*, *33*(2), 494–514. DOI: 10.1109/TNNLS.2021.3070843 PMID: 33900922

Kauffman, S. A. (1993). *The origins of order: Self-organization and selection in evolution*. Oxford University Press. DOI: 10.1093/oso/9780195079517.001.0001

Kerr, A. (2002). The Temple of Jesus' Body.

Kritzeck, J. (1964). *Peter the Venerable and Islam*. Princeton University Press. DOI: 10.1515/9781400875771

Krügel, S., Ostermaier, A., & Uhl, M. (2023). ChatGPT's inconsistent moral advice influences users' judgment. *Scientific Reports*, *13*(1), 4569. Advance online publication. DOI: 10.1038/s41598-023-31341-0 PMID: 37024502

Lane, J. E., & Shults, F. L. (2021). The computational science of religion. *Journal for the Cognitive Science of Religion*, *6*(1-2). Advance online publication. DOI: 10.1558/jcsr.38669

Lazer, D., Pentland, A., Adamic, L., Aral, S., Barabási, A.-L., Brewer, D., Christakis, N., Contractor, N., Fowler, J., Gutmann, M., Jebara, T., King, G., Macy, M., Roy, D., & Van Alstyne, M. (2009). Computational social science. *Science*, *323*(5915), 721–723. DOI: 10.1126/science.1167742 PMID: 19197046

LeCun, Y. (2022). *A path towards autonomous machine intelligence* (Version 0.9.2). OpenReview. https://openreview.net/forum?id=BZ5a1r-kVsf

Lee, Y. K., Lee, I., Shin, M., Bae, S., & Hahn, S. (2023). *Chain of empathy: Enhancing empathetic response of large language models based on psychotherapy models*. [Manuscript].

Lewis, P., Perez, E., Piktus, A., Petroni, F., Karpukhin, V., Goyal, N., & Kiela, D. (2020). Retrieval-augmented generation for knowledge-intensive nlp tasks. *Advances in Neural Information Processing Systems*, *33*, 9459–9474.

Lin, S., Hilton, J., & Evans, O. (2022). TruthfulQA: Measuring how models mimic human falsehoods. In *Proceedings of the 60th Annual Meeting of the Association for Computational Linguistics (*Volume 1*: Long Papers)* (pp. 3214–3252). Association for Computational Linguistics. DOI: 10.18653/v1/2022.acl-long.229

McClure, P. K. (2020). The buffered, technological self: Finding associations between Internet use and religiosity. *Social Compass*, *67*(3), 461–478. DOI: 10.1177/0037768620922128

McGrath, A. E. (2001–2003). *A scientific theology* (Vols. 1–3). Grand Rapids, MI: Eerdmans.

McGrath, A. E. (2017a). *Christian theology: An introduction* (6th ed.). Wiley-Blackwell.

McGrath, A. E. (2017b). *Theology: The basics* (4th ed.). Wiley-Blackwell. DOI: 10.1002/9781119158110

Meadows, D. H. (2008). *Thinking in systems: A primer*. Chelsea Green Publishing.

Mitchell, M. (2009). *Complexity: A guided tour*. Oxford University Press. DOI: 10.1093/oso/9780195124415.001.0001

Morreall, J. H., & Sonn, T. (2011). *The religion toolkit: A complete guide to religious studies*. Wiley-Blackwell.

Muhanna, E. (Ed.). (2016). *The digital humanities and Islamic & Middle East studies*. De Gruyter., DOI: 10.1515/9783110376517

Müller, F. M. (1873). *Introduction to the science of religion*. Longmans, Green.

Müller, F. M. (1878). *Lectures on the origin and growth of religion*. Longmans, Green.

Naqvi, A., & Munoz, M. (2020, March). *Rebuilding a company with artificial intelligence*. California Management Review. https://cmr.berkeley.edu/2020/03/rebuilding-a-company-with-ai/

Naqvi, J. (2022–2025). *Various lectures on YouTube* [Video series]. YouTube.

Naqvi, A., & Naqvi, N. (2026). *The Computational Quran for the AI Era: Selected Mathematical Models of Divine Guidance*. American Institute for AI and Quantum.

Nasr, S. H., Dagli, C. K., Dakake, M. M., Lumbard, J. E. B., & Rustom, M. (Eds.). (2015). *The Study Quran: A new translation and commentary*. HarperOne.

O'Gieblyn, M. (2021). *God, human, animal, machine: Technology, metaphor, and the search for meaning*. Doubleday.

Orsi, R. A. (2003). Is the study of lived religion irrelevant to the world we live in? *Journal for the Scientific Study of Religion*, *42*(2), 169–174. DOI: 10.1111/1468-5906.t01-1-00170

Otto, R. (1923). *The idea of the holy: An inquiry into the non-rational factor in the idea of the divine and its relation to the rational* (Harvey, J. W., Trans.). Oxford University Press.

Özcan, Ş. (2019). Ancient Origins of the History of Religions: Herodotus Example. *İlahiyat Tetkikleri Dergisi*, (51), 395-417.

Piotrowski, M. (2012). *Natural language processing for historical texts*. Morgan & Claypool Publishers., DOI: 10.1007/978-3-031-02146-6

Prigogine, I., & Stengers, I. (1984). *Order out of chaos: Man's new dialogue with nature*. Bantam Books.

Pyysiäinen, I. (2013). Cognitive science of religion: State-of-the-art. *Journal for the Cognitive Science of Religion*, *1*(1), 5–28. DOI: 10.1558/jcsr.v1i1.5

Rahman, F. (2009). *Major themes of the Qur'an* (2nd ed.). University of Chicago Press.

Raspanti, F., Beurer-Kellner, L., Fischer, M., & Vechev, M. (2025). Grammar-constrained decoding makes large language models better logical parsers. In *Proceedings of the 63rd Annual Meeting of the Association for Computational Linguistics: Industry Track*. Association for Computational Linguistics. DOI: 10.18653/v1/2025.acl-industry.34

Ratican, J., & Hutson, J. (2023). The six emotional dimension (6DE) model: A multidimensional approach to analyzing human emotions and unlocking the potential of emotionally intelligent artificial intelligence via large language models. *1*(1), 44–52.

Russell, S. (2019). *Human compatible: Artificial intelligence and the problem of control*. Viking.

Saeed, A. (2006). *Interpreting the Qur'an: Towards a contemporary approach*. Routledge., DOI: 10.4324/9780203016770

Schreibman, S., Siemens, R., & Unsworth, J. (Eds.). (2004). *A companion to digital humanities*. Blackwell.

Sharpe, E. J. (1986). *Comparative religion: A history* (2nd ed.). Duckworth.

Singler, B. (2024). *Religion and artificial intelligence: An introduction*. Routledge., DOI: 10.4324/9781003256113

Smart, N. (1998). *The world's religions* (2nd ed.). Cambridge University Press.

Stacey, R. D. (2011). *Strategic management and organisational dynamics: The challenge of complexity to ways of thinking about organisations* (6th ed.). Pearson Education Limited.

Tylor, E. B. (1871). *Primitive culture* (Vol. 1–2). John Murray.

von Bertalanffy, L. (1968). *General system theory: Foundations, development, applications*. George Braziller.

von Neumann, J. (1955). *Mathematical foundations of quantum mechanics* (Beyer, R. T., Trans.). Princeton University Press.

Wang, X., Li, X., Yin, Z., Wu, Y., & Liu, J. (2023). *Emotional intelligence of large language models*. DOI: 10.1177/18344909231213958

Wiener, N. (1948). *Cybernetics: Or control and communication in the animal and the machine*. MIT Press.

Zhou, J., Hu, M., Li, J., Zhang, X., Wu, X., King, I., & Meng, H. (2022). *Rethinking machine ethics: Can LLMs perform moral reasoning through the lens of moral theories?* [Manuscript]. Top of Form Bottom of Form

About the Authors

Nurullah Naqvi is the chief Technology Officer and President of the American Institute of Artificial Intelligence, a research institute and government contractor specializing in advanced AI and machine learning solutions. Leads the development and deployment of GenAI-integrated systems, including a next-generation online learning platform and mission-aligned AI solutions for national defense. Recent technical leadership includes architecting and managing transformer-based multi-task learning systems for radar signal characterization, vector symbolic architectures for signal analysis, and quantum-inspired algorithm development. Previously oversaw the creation of a graph-based machine learning platform for B2B intelligence supporting federal contracting operations. Graduate education in Applied and Computational Mathematics at Johns Hopkins University, with a focus on AI, ML, quantum computing, and scientific computing. Deep hands-on expertise in Python, Java, and MATLAB, applied across projects in quantum circuit design, encryption algorithms, natural language processing, topological data analysis, and numerical modeling. Combines mathematical rigor, systems thinking, and executive leadership to drive innovative GenAI product development for mission-critical domains.

Index

Q

R

S

T

www.ingramcontent.com/pod-product-compliance
Lightning Source LLC
LaVergne TN
LVHW081524300926
849161LV00006B/1293